Finite Element
Analysis Concepts
Via SolidWorks

Finite Element Analysis Concepts
Via SolidWorks

John Edward Akin

Rice University, USA

World Scientific

NEW JERSEY · LONDON · SINGAPORE · BEIJING · SHANGHAI · HONG KONG · TAIPEI · CHENNAI

Published by

World Scientific Publishing Co. Pte. Ltd.

5 Toh Tuck Link, Singapore 596224

USA office: 27 Warren Street, Suite 401-402, Hackensack, NJ 07601

UK office: 57 Shelton Street, Covent Garden, London WC2H 9HE

British Library Cataloguing-in-Publication Data
A catalogue record for this book is available from the British Library.

FINITE ELEMENT ANALYSIS CONCEPTS
Via SolidWorks

ISBN-13 978-981-4313-01-8
ISBN-10 981-4313-01-7

Typeset by Stallion Press
Email: enquiries@stallionpress.com

Printed in Singapore by Mainland Press Pte Ltd

Preface

This book is intended for beginners who must utilize finite element analysis (FEA) methods, but have not yet had a course in finite element theory. The emphasis is on the engineering reasons for conducting the sequence of stages necessary to complete a valid finite element analysis. For completeness, it has been necessary to select specific software to illustrate the various stages. Most commercial solid modeling and finite element analysis systems are very similar, and the overlap in their capabilities is probably 90% or more. Systems with more advanced or specialized abilities are usually more difficult to utilize, and have a significantly longer learning curve. In this case, the SolidWorks system (release 2010) has been selected due to its short learning curve and ability to execute the most commonly needed finite element analyses. The author wishes to thank the SolidWorks Corporation for permission to reproduce some of their online help file or demonstration figures for use in this text.

This book is based on my forty years of teaching finite element analysis, applying it in consulting applications and programming FEA. Historically, the typical initial users of FEA were often graduate level engineers well educated in the prerequisite knowledge. Most had completed courses covering statics, dynamics, free body diagrams, material properties, stress analysis, heat transfer, vibrations, etc. The slow computers of the time also required that they understand numerical analysis methods, and that they often programmed their own finite element systems. The good news is that, as implied by Moore's Law, today's computers have grown so powerful that one does not have to have the previously prerequisite knowledge to be able to easily build solid models of complex geometries and then to apply all sorts of FEA to those models. That is also the bad news. Users no longer must have degrees in engineering (or applied mathematics) or even be exposed to the basic knowledge about materials and their stress or thermal responses. Indeed, systems like the SolidWorks software are currently being utilized by high school students in several regions.

Those realities mean that often what seems like computer aided design has become computer aided errors, or computer aided stupidity. To avoid such problems this book will hopefully identify the engineering reasons for selecting various capabilities, rather than just the sequence of icon picks that yield the pretty pictures that often hide misleading or erroneous results or assumptions. To support that educational process, the basic concepts or definitions of desirable prerequisite knowledge will be covered as needed. Having stated those goals, the reader of this text is strongly encouraged to complete a course in basic finite element theory. It can be dangerous to utilize tools when you do not understand their fundamental abilities and limitations.

Modern commercial finite element systems typically include millions of lines of source code. They are continuously being modified. Therefore, they are likely to contain some errors that could, on rare occasions, affect your analysis results. A good engineer always tries to check any analysis results. Thus, as space permits, I have also included typical examples of attempts to validate sample finite element analysis results.

J. Ed Akin, PhD, PE
Professor
Rice University
akin@rice.edu
December 2009

Contents

1

Finite Element Analysis Methods

1.1. Introduction

The finite element method (FEM) rapidly grew as the most useful numerical analysis tool for engineers and applied mathematicians because of it natural benefits over prior approaches. The main advantages are that it can be applied to arbitrary shapes in any number of dimensions. The shape can be made of any number of materials. The material properties can be non-homogeneous (depend on location) and/or anisotropic (depend on direction). The way that the shape is supported (also called fixtures or restraints) can be quite general, as can the applied sources (forces, pressures, heat flux, etc.). The FEM provides a standard process for converting governing energy principles or governing differential equations in to a system of matrix equations to be solved for an approximate solution. For linear problems, such solutions can be very accurate and quickly obtained. Having obtained an approximate solution, the FEM provides additional standard procedures for follow up calculations (post-processing), such as determining the integral of the solution, or its derivatives at various points in the shape. The post-processing also yields impressive color displays, or graphs, of the solution and its related information. Today, a second post-processing of the recovered derivatives can yield error estimates that show where the study needs improvement. Indeed, adaptive procedures allow automatic corrections and re-solutions to reach a user specified level of accuracy.

1

However, very accurate and pretty solutions of models that are based on errors or incorrect assumptions are still wrong.

When the FEM is applied to a specific field of analysis (like stress analysis, thermal analysis, or vibration analysis) it is often referred to as finite element analysis (FEA). An FEA is the most common tool for stress and structural analysis. Various fields of study are often related. For example, distributions of non-uniform temperatures induce non-obvious loading conditions on solid structural members. Thus, it is common to conduct a thermal FEA to obtain temperature results that in turn become input data for a stress FEA. FEA can also receive input data from other tools like motion (kinetics) analysis systems and computation fluid dynamics systems.

1.2. Basic Integral Formulations

The basic concept behind the FEM is to replace any complex shape with the union (or summation) of a large number of very simple shapes (like triangles) that are combined to correctly model the original part. The smaller simpler shapes are called finite elements because each one occupies a small but finite sub-domain of the original part. They contrast to the infinitesimally small or differential elements used for centuries to derive differential equations. To give a very simple example of this dividing and summing process, consider calculating the area of the arbitrary shape shown in Figure 1.1 (left).

If you knew the equations of the bounding curves you, in theory, could integrate them to obtain the enclosed area. Alternatively, you could split the area into an enclosed set of triangles (cover the shape

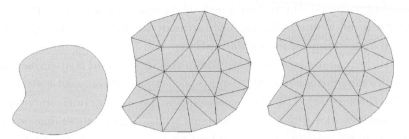

Fig. 1.1.   An area crudely meshed with linear and quadratic triangles.

with a mesh) and sum the areas of the individual triangles:

$$A = \sum_{e=1}^{n} A^e = \sum_{e=1}^{n} \int_{A^e} dA.$$

Now, you have some choices for the type of triangles. You could pick straight sided (linear) triangles, or quadratic triangles (with edges that are parabolas), or cubic triangles, etc. The area of a straight-sided triangle is a simple algebraic expression. Number the three vertices in a counter-clockwise order, then the area is $A^e = [x_1(y_2 - y_3) + x_2(y_3 - y_1) + x_3(y_1 - y_2)]/2$ and its centroid is located at

$$x_{cg}^e = \frac{x_1 + x_2 + x_3}{3}, \quad y_{cg}^e = \frac{y_1 + y_2 + y_3}{3}.$$

Similar expressions give the moment of inertia components. Thus, you just have to extract (gather) the element vertices coordinates from the mesh data in order to compute the area of a straight sided triangle. The area of a curved triangle is also relatively easy to calculate by numerical integration, but is computationally more expensive to obtain than that for the linear triangle. The first two triangle mesh choices are shown in Figure 1.1 for a large element size. Clearly, the simple straight-sided triangular mesh (on the left) approximates the area very closely, but at the same time introduces geometric errors along a curved boundary. The boundary geometric error in a linear triangle mesh results from replacing a boundary curve by a series of straight line segments. That geometric boundary error can be reduced to any desired level by increasing the number of linear triangles. But that decision increases the number of calculations and makes you trade off geometric accuracy versus the total number of required area calculations and summations.

Area is a scalar, so it makes sense to be able to simply sum its parts to determine the total value, as shown above. Other physical quantities, like kinetic energy, strain energy and mechanical work, can be summed in the same fashion. Indeed, the very first applications of FEA to structures was based on minimizing the energy stored is a linear elastic material. The FEM always involves some type of governing integral statement which is converted to a matrix system by assuming how items vary within a typical element. That integration is also converted to the sum of the integrals over each element in the mesh. Even if you start with a governing differential equation, it gets converted to an equivalent integral formulation by one of the methods of weighted residuals (MWR). The two most common weighted residual methods, for FEA, are the Galerkin method and the Method of Least Squares.

The development of the necessary matrix relations will be covered in more detail later. For now the matrix representation of the kinetic energy of a part is presented, for the straight sided triangle element, as an example. Recall that the kinetic energy of a mass particle is $KE = mv^2/2$, where m is the mass and v is its velocity. The kinetic energy of the planar body, of thickness t, in Figure 1.1 is obtained by integrating over the differential masses

$$KE = \frac{1}{2} \int v^2 dm = \frac{1}{2} \int v^2 \rho dV = \frac{t}{2} \int v^2 \rho dA$$

and it can be obtained by summing the element integrals

$$KE = \sum_{e=1}^{n} KE^e = \frac{t}{2} \sum_{e=1}^{n} \int_{A^e} v^2 \rho dA.$$

Assume that the velocity of all mesh points is known. Then you can extract the velocity of the three vertices of each linear triangle element. In FEA you define the velocity of any point inside the element by interpolating between the element's vertex values. Denote the interpolation function of node j by $N(x,y)_j$. Then the velocity is

$$v(x,y) = N(x,y)_1 \cdot v_1 + N(x,y)_2 \cdot v_2 + N(x,y)_3 \cdot v_3$$

$$= \sum_{j=1}^{3} N(x,y)_j \cdot v_j.$$

This linear spatial interpolation is usually written in a matrix notation

$$v(x,y) = [N(x,y)]\{v^e\} = \{v^e\}^T [N(x,y)]^T$$

so the element kinetic energy is a matrix integral

$$KE^e = \frac{1}{2} \int_{A^e} \{v^e\}^T [N(x,y)]^T \rho [N(x,y)] \{v^e\} t \, dA$$

but the nodal velocities are known constants (or functions of time alone) that move outside the integral

$$KE^e = \frac{1}{2} \{v^e\}^T \int_{A^e} [N(x,y)]^T \rho [N(x,y)] t \, dA \{v^e\}$$

so that the remaining square matrix integral is called the mass matrix, $[m^e]$:

$$KE^e = \frac{1}{2} \{v^e\}^T [m^e] \{v^e\}, \quad [m^e] = \int_{A^e} [N(x,y)]^T \rho [N(x,y)] t \, dA.$$

This result shows the similarity to a particle in that the body's kinetic energy is half the product of a mass matrix and the square (pre- and post-multiplication) of the nodal velocities. For the straight two-node bar element of length L^e and cross-sectional area A^e, the mass matrix is

$$[m^e] = \frac{\rho L^e A^e}{6} \begin{bmatrix} 2 & 1 \\ 1 & 2 \end{bmatrix}.$$

The total element mass, $m = \rho L^e A^e$, is present but somehow distributed (coupled) between its two nodes, and their velocities, $\{v^e\}^T = [v_1^e \quad v_2^e]$. If the two end velocities are the same (it is moving as a rigid body such that $v_1^e = v_2^e = v$) then the matrix products yield the expected scalar answer, $KE^e = mv^2/2$. The point of this illustration is to show that any FEA converts scalar integrals to matrix expressions by assuming a spatial interpolation between the nodes of a typical element for items of interest, such as positions, displacements, velocities, or temperatures. Those spatial interpolations also define the spatial gradients that occur in most finite element integral forms.

1.3. Gather and Scatter Operators

An integral evaluation for an FEA requires a mesh. Typically it is a triangular mesh for surfaces and a tetrahedral mesh for solids.

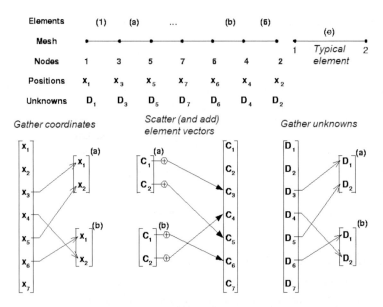

Fig. 1.2.   Matrix gather and scatter operations.

The result of a finite element mesh generation creates at least two data sets. The first (nodal set) is the numbered list of all the generated vertices along with their spatial coordinates. The second (element set) is the numbered set of elements along with the list of element vertex numbers to which it is connected. This is called the element connectivity list. Figure 1.2 illustrates such data for linear elements.

The connectivity list is the critical data that allows the FEA calculations to be automated. Any FEA uses operations that involve the specific node (vertex) numbers of a single element. The two operations are usually called gather and scatter (or assembly) operations. The gather operation is used to simply bring known nodal data in the full mesh back to a single element.

The coordinates and velocities used in the above element integrals were assumed to be stored with the mesh nodal data. While the mesh may have a huge number of nodes, each linear triangle element only had three nodes. The gather operator utilized the element connection list to extract the data for the current element in the summation to extract its three nodal velocities.

The reverse of a gather operation is the scatter or assembly operation. It is a partial summation of element data to the matrices associated with the mesh data. A scatter takes something associated with the local nodes of an element and adds them to the corresponding matrix item at the full mesh level. Scatter, or assembly, operations fill the entries in the matrix equations that must be solved for the problem unknowns. These two common operations are sketched graphically in Figure 1.2. The required element connectivity data (the two nodes on each linear line element) are displayed in the third line of that figure. Gather and scatter will be illustrated in detail in the section on compound elastic bars.

1.4. Geometric Boundary Errors

You may think that the geometric boundary error cited for the linear triangles is eliminated by choosing to use the mesh of curved quadratic triangles (on the right of Figure 1.1). The parabola segments pass through three points lying exactly on the boundary curve, but can degenerate to straight lines in the interior. (To speed plotting of small elements, most systems draw all the parabolas as two straight line segments, as on the right in Figure 1.1.) Thus, the boundary shape error is indeed reduced, at the expense of more complicated area calculations, but it is not eliminated. Some geometric error remains because most engineering curves are circular arcs, splines, or nurbs (non-uniform rational B-splines) and thus are not matched by a parabola. The most common way to reduce mesh geometric error is to simply use many smaller elements. The default element choice in SolidWorks (SW) Simulation is the quadratic element. Other systems offer a wider range of edge polynomial degree (e.g., cubic), as well as other shapes like quadrilaterals or rectangles. In three-dimensional solid applications some systems offer dozens of choices for the edge degree polynomial order, and shapes including hexahedral, wedges, and tetrahedral elements. Hexahedral elements are generally more accurate, but can be more challenging to mesh automatically. Tetrahedral elements can match hexahedral element performance by using more (smaller) elements, and tetrahedral

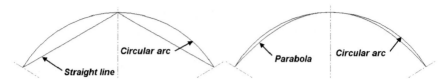

Fig. 1.3.   Linear or parabolic elements never match circular shapes.

elements are much easier to mesh automatically. SW Simulation uses only tetrahedral elements for solid studies. An example of the small two-dimensional geometric boundary error due to different curved shapes is seen in Figure 1.3 where a arc and a parabola pass through the same three points.

1.5. Stages of Analysis and Their Uncertainties

An FEA always involves a number of uncertainties that impact the accuracy or reliability of each stage of an FEA and its results. The book, *Building Better Products with Finite Element Analysis* by Adams and Askenazi [1] gives an outstanding detailed description of most of the real-world uncertainties associated with solid mechanics FEA. All engineers conducting stress studies should read it. That book also points out how poor solid modeling skills can adversely affect the ability to construct meshes for any type of FEA. Here, the most important FEA uncertainties are highlighted.

The typical stages of an FEA study are listed below:

(1) Construct the part(s) in a solid modeler. It is surprisingly easy to accidentally build flawed models with tiny lines, tiny surfaces or tiny interior voids. The part will look fine, except with extreme zooms, but it may fail to mesh. Most systems have checking routines that can find and repair such problems before you move on to an FEA study. Sometimes you may have to export a part, and then import it back with a new name because imported parts are usually subjected to more time consuming checks than "native" parts. When multiple parts form an assembly, always mesh and study the individual parts before studying the assembly. Try to plan ahead and introduce

split lines into the part to aid in mating assemblies and to locate load regions and restraint (or fixture or support) regions. Today, construction of a part is probably the most reliable stage of any study.

(2) Defeature the solid part model for meshing. The solid part may contain features, like a raised logo, that are not necessary to manufacture the part, or required for an accurate analysis study. They can be omitted from the solid used in the analysis study. That is a relative easy operation supported by most solid modelers (such as the "suppress" option in SW) to help make smaller and faster meshes. However, it has the potential for introducing serious, if not fatal, errors in a following engineering study. This is a reliable modeling process, but its application requires engineering judgment. For example, removing small radius interior fillets can greatly reduces the number of elements and simplifies the mesh generation. But, that creates sharp reentrant corners that can yield false infinite stresses. Those false high stress regions may cause you to overlook other areas of true high stress levels. Small holes lead to many small elements (and long run times). They also cause stress concentrations that raise the local stress levels by a factor of three or more. The decision to defeature them depends on where they are located in the part. If they lie in a high stress region you must keep them. But defeaturing them is allowed if you know they occur in a low stress region. Such decisions are complicated because most parts have multiple possible loading conditions and a low stress region for one load case may become a high stress region for another load case.

(3) Combine multiple parts into an assembly. Again, this is well automated and reliable from the geometric point of view and assemblies "look" as expected. However, geometric mating of part interfaces is very different for defining their physical (displacement, or temperature) mating. The physical mating choices are often unclear and the engineer may have to make a range of assumptions, study each, and determine the worst case result. Having to use physical contacts makes the linear

problem require iterative solutions that take a long time to run and might fail to converge.

(4) Select the element type. Some FEA systems have a huge number of available element types (with underlying theoretical restrictions). The SolidWorks system has only the fundamental types of elements. Namely, truss elements (bars), frame elements (beams), thin shells (or flat plates), thick shells, and solids. The SW simulation system selects the element type (beginning in 2009) based on the shape of the part. The user is allowed to covert a non-solid element region to a solid element region, and vice versa. Knowing which class of element will give a more accurate or faster solution requires training in finite element theory. At times a second element type study is used to help validate a study based on a different element type.

(5) Mesh the part(s) or assembly, remembering that the mesh solid may not be the same as the part solid. A general rule in an FEA is that your computer never has enough speed or memory. Sooner or later you will find a study that you cannot execute. Often that means you must utilize a crude mesh (or at least crude in some region) and/or invoke the use of symmetry or anti-symmetry conditions. Local solution errors in a study are proportional to the product of the local element size and the gradient of the secondary variables (i.e., gradient of stress or heat flux). Therefore, you exercise mesh control to place small elements where your engineering judgment estimates high stress (or flux) regions, as well as large elements in low stress regions. The local solution error also depends on the relative sizes of adjacent elements. You do not want skinny elements adjacent to big ones. Thus, automatic mesh generators have options to gradually vary adjacent element sizes from smallest to biggest.

The solid model sent to the mesh generator frequently should have load or restraint (fixture) regions formed by split lines, even if such splits are not needed for manufacturing the parts. The mesh typically should have refinements at source or load regions and support regions.

A mesh must look like the part, but that is not sufficient for a correct study. A single layer of elements filling a part region is almost never enough. If the region is curved, or subjected to bending, you want at least three layers of quadratic elements, but five is a desirable lower limit. For linear elements you at least double those numbers.

Most engineers do not have access to the source code of their automatic mesh generator. When the mesher fails, you frequently do not know why it failed or what to do about it. Often you have to re-try the mesh generation with very large element sizes in hopes of getting some mesh results that can give hints as to why other attempts failed. The meshing of assemblies often fails. Usually the mesher runs out of memory because one or more parts had a very small, often unseen, feature that causes a huge number of tiny elements to be created (see the end of this chapter). You should always attempt to mesh each individual part to spot such problems before you attempt to mesh them as a member of an assembly.

Automatic meshing, with mesh controls, is usually simple and fast today. However, it is only as reliable as the modified part or assembly supplied to it. Distorted elements usually do not develop in automatic mesh generators, due to empirical rules for avoiding them. However, distorted elements locations can usually be plotted. If they are in regions of low gradients you can usually accept them.

You should also note that studies involving natural frequencies are influenced most by the distribution of the mass of the part. Thus, they can still give accurate results with meshes that are much cruder than those that would be acceptable for stress or thermal studies.

(6) Assign a linear material to each part. Modern FEA systems have a material library containing the "linear" mechanical, thermal, and/or fluid properties of most standardized materials. They also allow the user to define custom properties. The property values in such tables are often misinterpreted to be more accurate and reliable than they actually are. The reported

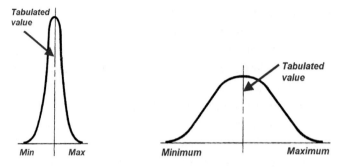

Fig. 1.4.   Typical distributions of properties of steel (left) and cast iron.

property values are accepted average values taken from many tests. Rarely are there any data about the distribution of test results, or what standard deviation was associated with the tests. Most tests yield results that follow a "bell shaped" curve distribution, or a similar skewed curve. The tests for stainless steel tend to have narrow distributions, like that on the left in Figure 1.4, while the results for cast iron have wider distributions.

When you accept a tabulated property value as a single number to be used in the FEA calculation remember it actually has a probability distribution associated with it. You need to assign a contribution to the total factor of safety to allow for variations from the tabulated property value.

The values of properties found in a material table can appear more or less accurate depending on the units selected. That is an illusion often caused by converting one set of units to another, but not truncating the result to the same number of significant figures available in the actual test units. For example, the elastic modulus of one steel is tabulated from the original test as 210 MPa, but when displayed in other units it shows as 30,457,924.92 psi. Which one do you believe to be the experimental accurate value; the 3 digit value or the 10 digit one? The answer affects how you should view and report stress results. The axial stress in a bar is equal to the elastic modulus times the strain, $\sigma = E\varepsilon$. Thus, if E is only known to three or four significant figures then the reported stress result should have no more significant figures.

Material data are usually more reliable than the loading values (considered next), but less accurate that the model or mesh geometries.

(7) Select regions of the part(s) to be loaded and assign load levels and load types to each region. In mathematical terminology, load or flux conditions on a boundary region are called Neumann boundary conditions, or non-essential conditions. The geometric regions can be points (in theory), lines, surfaces, or volumes. If they are not existing features of the part, then you should apply split lines to the part to create them before activating the mesh generator. Point forces, or heat sources, are common in undergraduate studies, but in an FEA they cause false infinite stresses, or heat flux. If you include them do not be mislead by the high local values. Refining the mesh does not help since the smallest element still reports near infinite values.

In reality, point loads are better modeled as a total force, or pressure, acting over a small area formed by prior split lines. Saint Venant's Principle states that two different, but statically equivalent, force systems acting on a small portion of the surface of a body produce the same stress distributions at distance large in comparison with the linear dimensions of the portion where the forces act. That also implies that concentrated sources quickly become re-distributed, as seen in Figure 1.5. There a single axial forced on the right end has been replaced by a small region of constant pressure. The other end stresses are essentially zero. Within the distance of about one depth the

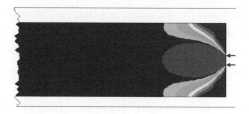

Fig. 1.5.   St. Venant's principle: local effects quickly die out.

axial stress has re-distributed to an essentially constant value in the remainder of the part.

In undergraduate statics and dynamics courses engineers are taught to think in terms of point forces and couples. Solid elements do not accept pure couples as loads, but statically equivalent pressures can be applied to solids and yield the correct stresses. Indeed, a couple at a point is almost impossible to create, so the distribution of pressures is probably more like the true situation.

The magnitudes of applied loads are often guesses, or specified by a governing design standard. For example, consider a wind load. A building standard may quote a pressure to be applied for a given wind speed. But, how well do you know the wind speed that might actually be exerted on the structure? Again, there probably is some type of "bell curve" around the expected average speed. You need to assign a contribution to the total factor of safety to allow for variations in the uncertainty of the load value or actual spatial distribution of applied loads.

Loading data are usually less accurate than the material data, but much more accurate than the restraint or supporting conditions considered next.

(8) Determine (or more likely assume) how the model interacts with the surroundings not included in your model. These are the restraint (support, or fixture) regions. In mathematical terminology, these are called the essential boundary conditions, or Dirichlet boundary conditions. You cannot afford to model everything interacting with a part. For many decades engineers have developed simplified concepts to approximate surroundings adjacent to a model to simplify hand calculations. They include roller supports, smooth pins, cantilevered (encastre, or fixed) supports, straight cable attachments, etc. Those concepts are often carried over to FEA approaches and can over simplify the true support nature and lead to very large errors in the results.

The choice of restraints (fixations, supports) for a model is surprisingly difficult and is often the least reliable decision made by the engineer. Small changes in the supports can cause large changes in the results. It is wise to try to investigate a number of likely or possible support conditions in different studies. When in doubt, try to include more of the surrounding support material and apply assumed support conditions to those portions at a greater distance from critical part features.

You need to assign a contribution to the total factor of safety to allow for variations in the uncertainty of how or where the actual support conditions occur. That is especially true for buckling studies.

(9) Solve the linear system of equations, or the eigenvalue problem. With today's numerical algorithms the solution of the algebraic system or eigen-system is usually quite reliable. It is possible to cause ill-conditioned systems (large condition number) with meshes having bad aspect ratios, or large elements adjacent to small ones, but that is unlikely to happen with automatic mesh generators.

(10) Check the results. Are the reactions at the supports equal and opposite to the sources you thought that you applied? Are the results consistent with the assumed linear behavior? The engineering definition of a problem with large displacements is one where the maximum displacement is more than half the smallest geometric thickness of the part. The internal definition is a displacement field that significantly changes the volume of an element. That implies the element geometric shape noticeably changed from the starting shape, and that the shape needs to be updated in a series of much smaller shape changes. Are the displacements big enough to require re-solution with large displacement iterations turned on? Have you validated the results with an analytic approximation, or different type of finite element? Engineering judgments are required.

(11) Post-process the solution for secondary variables. For structural studies you generally wish to document the deflections, reactions and stresses. For thermal studies you display the temperatures, heat flux vectors and reaction heat flows. With natural frequency models you show (or animate) a few mode

shapes. In graphical displays, you can control the number of contours employed, as well as their maximum and minimum ranges. The latter is important if you want to compare two designs on a single page. Limit the number of digits shown on the contour scale to be consistent with the material modulus (or conductivity, etc.). Color contour plots often do not reproduce well, but graphs do, so learn to use them in your documentation.

(12) Determine (or more likely assume) what failure criterion applies to your study. This stage involves assumptions about how a structural material might fail. There are a number of theories. Most are based on stress values or distortional energy levels, but a few depend on strain values. If you know that one has been accepted for your selected material then use that one (as a contribution to the overall factor of safety). Otherwise, you should evaluate more than one theory and see which is the worst case. Also keep in mind that loading or support uncertainties can lead to a range of stress levels, and variations in material properties affect the strength and unexpected failures can occur if those types of distributions happen to intersect, as sketched in Figure 1.6.

(13) Optionally, post-process the secondary variables to measure the theoretical error in the study, and adaptively correct the solution. This converges to an accurate solution to the problem input, but perhaps not to the problem to be solved. Accurate garbage is still garbage.

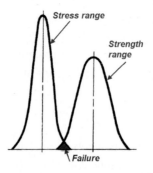

Fig. 1.6.   Overlap of stresses and material strengths can cause failure.

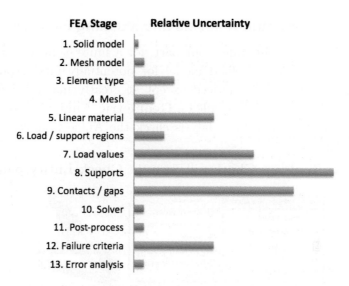

Fig. 1.7. Relative uncertainty of major modeling stages.

(14) Document, report, and file the study. The part shape, mesh, and results should be reported in image form. Assumptions on which the study was based should be clearly stated, and hopefully confirmed. The documentation should contain an independent validation calculation, or two, from an analytical approximation or an FEA based on a totally different element type. Try to address the relative uncertainties of the main analysis stages, as summarized in Figure 1.7.

Technical communication and documentation is always important. In America, engineers are supposed to retain their calculations for at least seven years. Will your report be clear and helpful if you have to defend it years later? Paper hardcopies are the most reliable for long term storage. (Can you read the electronic media you used five years ago?)

You usually assume that the materials are linear. If not (creeping, hyperelastic, inelastic, plastic, viscoelastic, etc.), define the appropriate material data and the nonlinear equations to be solved. Then the matrix system becomes non-linear. Your original results check may lead you to conclude that the problem is actually an iterative one due to large displacements, or the need to insert physical contact interfaces.

1.6. Part Geometric Analysis and Meshing Failures

Before attempting meshing your part, for a finite element analysis, you should check your solid model for potentially fatal geometric flaws that may not be noticed except at greatly magnified views. Within SolidWorks this is called a **Geometric Analysis**. To utilize that feature, a geometric analysis check the Angle_Connector part will be outlined:

(1) Select **Tools → Check** will open the **Check Entity panel**.

(2) In that panel check the boxes for most entities, select **Check**.
(3) Highlight each item in the **Result List**. As you scroll down the Result list the short edge location on the part is illustrated by a yellow arrow. Either the feature needs to be eliminated, or the mesh will need to be finer.

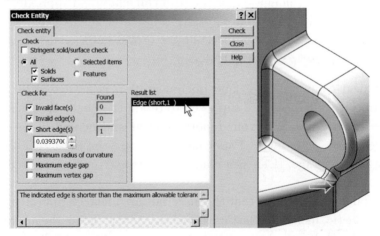

(4) To consider a potential mesh refinement you should determine the size of the small feature. Use **Tools → Measure** to open up the **Measure panel**. Select the **XYZ** option, click on a edge of the feature to see its length.

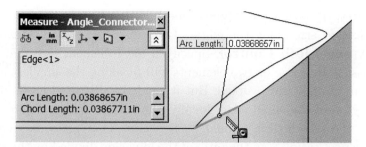

(5) Attempt to create a mesh: **Mesh** → **Create Mesh**. As expected, that process fails and a failure diagnostic message appears:

(6) Right click on **Mesh** to open the **Failure Diagnostics panel**. Scroll down the lists of faces or edges that caused the meshing failure. In this case, there is a highly distorted surface that formed with the fillets. Sometimes this type of surface can be removed by suppressing the fillets, or by simply building the fillets in a different order. Sometimes the surface can split by inserting split lines to make more manageable regions. Repairing the surface is better that having to *try* to control the mesh.

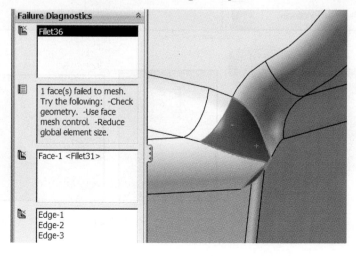

(7) First, try to get some type of mesh output by specifying a small element size along the edges of the distorted region **Mesh →**
Apply Mesh Control. Specify a local element size that will assure that one or two elements will fit along the smallest edge. Surprisingly, this worked. But it yielded a distorted mesh in the region of the small edge. Ideally, the surface triangles (one face of the tetrahedron) would be isosceles. That gives an element "aspect ratio" (say the ratio of the long side divided by the short one) of unity. Here the triangles are curved. A few are also badly distorted and not desirable for analysis if they are in an expected high stress region.

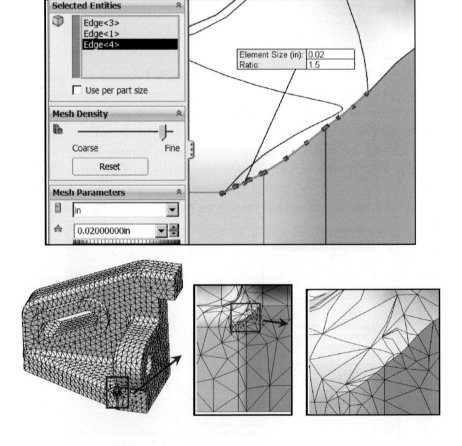

One measure of the quality of an element is its aspect ratio. Think of that as the ratio of the diameter of the enclosing sphere to the diameter of the enclosed sphere. Alternately, use the ratio of the longest element edge length to its shortest. An ideal aspect ratio should be near unity. Check the mesh quality by looking at a plot of the aspect ratio of the elements. Select **Mesh → Create Mesh Plot → Aspect Ratio**.

Try to improve this mesh by removing the bad surface, or subdividing it into two regions. At the narrow region, insert a **split line** that avoids very small intersection angles with both curves.

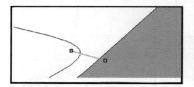 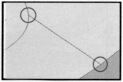

The small slender partition will need very small elements, but the larger partition can have larger ones. Especially if you use the transition control ratio to give five or more growth layers at an enlargement ratio of about 1.2 instead of the default value of 1.5. Use **Mesh → Apply Mesh Control** to specify element sides of *0.02* and *0.05 inches*, respectively in the **Mesh Control panel**. They give a much better mesh in this region.

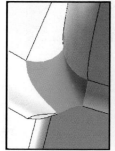 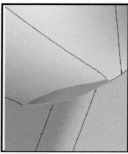 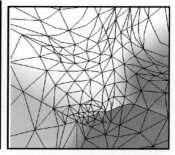

Another part, the Five_Hole_Link, shows a tangency that gives very bad element aspect ratios. A common cause of failure in mesh generation is to have two solid regions or two joining surfaces meet at a near zero angle. That often happens in practice and requires

 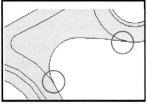

intervention to be able to create a mesh for analysis. If a tangency condition is really required in the part, then you must force smaller element sizes there via the **Mesh Control** option.

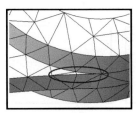 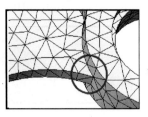

If the part can be modified to avoid the tangency, then meshing becomes much easier. That is illustrated below where it was feasible to avoid a tangency requirement in this application.

SolidWorks Simulation Overview

2.1. Simulation Study Capabilities

The SW Simulation software offers several types of linear studies including:

Buckling: A buckling study calculates a load factor multiplier for axial loads to predict when the actual loads will cause sudden, large catastrophic transverse displacements. Slender structures subject to mainly axial loads can fail due to buckling at load levels far lower than those required to cause material failure.

Drop Test: Drop test studies evaluate the impact effect of dropping the design on a rigid floor. You can specify the dropping distance or the velocity at the time of impact in addition to gravity. The model supplies the mass, M, damping, C, and stiffness matrices, K, in terms of the displacements, $u(t)$, and forces, F. The program solves a dynamic problem $M\ddot{u} + C\dot{u} + Ku(t) = F(t)$ as a function of time using explicit time integration methods. After the analysis is completed, you can plot and graph the time history of the displacements, velocities, accelerations, strains, and stresses.

Dynamic Analysis: These types of study assume that the materials are linear and that the loadings are either time dependent, frequency dependent or defined by limiting spectra. Mass and inertia effects are included and damping is available. The options in SW Simulation are Drop Test (also known as

Direct Time History Analysis), Modal Time History Analysis (Mode Superposition Analysis), Harmonic Analysis (Harmonic Response Analysis, and Random Vibration Analysis (Response Spectra Analysis). The last three analysis types require a Frequency Analysis to be completed to supply the eigenvalues (natural frequencies) and eigenvectors (mode shapes) needed as inputs.

Fatigue: Fatigue studies evaluate the consumed life of an object based on a very large number of fatigue events (cycles). Repeated loading weakens materials over time even when the induced stresses are low. The number of cycles required for failure depends on the material and the stress fluctuations. Those data are provided by the material S-N curve, which depicts the number of cycles that cause failure for different stress levels.

Frequency: A body tends to vibrate at natural, or resonant, frequencies. For each natural frequency, the body takes a certain shape called the mode shape. Frequency analysis calculates the natural frequencies and their associated mode shapes. The mode shapes can be animated for each selected frequency.

Harmonic Analysis: Harmonic response analysis is a steady state solution due to harmonic loads of known amplitude and frequency. That is $F(t) = F_0 e^{j\omega t}$, and $u(t) = u_0 e^{j\omega t}$ and the linear system becomes $(-\omega^2 M + j\omega C + K)u_0 = F_0$ which is solved for u_0 and its associated strains and stresses.

Optimization: Optimization studies automate the search for a local optimum design based on an initial geometric design and analysis state. Optimization studies require the definition of an objective, design variables, and behavior constraints.

Modal Time History: This solves the matrix equations of motion, $M\ddot{u} + C\dot{u} + Ku(t) = F(t)$, where the assembled loadings, $F(t)$, are functions of time. The displacement unknowns, $u(t)$, are converted through a modal transformation, $u = \psi q$, to generalized DOF, $q(t)$. This gives a diagonal matrix system

$\ddot{q} + \lceil 2\omega_j \xi_j \rfloor \dot{q} + \lceil \omega_k^2 \rfloor q(t) = \psi^T F(t)$ which is integrated analytically for a typical time step. When the time history is completed, the physical displacements, $u(t)$, are recovered. The strains and stresses are obtained from the displacements at each time.

Pressure Vessel Design: The results of multiple static studies are combined with the desired load factors. This study combines the results algebraically using a linear combination or the square root of the sum of the squares.

Random Vibration: The loads are described statistically by power spectral density (PSD) functions. After running the study, you can plot root-mean-square (RMS) values, or PSD results of stresses, displacements, velocities, etc. at a specific frequency or graph results at specific locations versus frequency values.

Static: Static (or Stress) studies calculate displacements, reaction forces, strains, stresses, failure criterion, factor of safety, and error estimates. Available loading conditions include point, line, surface, acceleration (volume) and thermal loads are available. Elastic orthotropic materials are available.

Thermal: Thermal studies calculate temperatures, temperature gradients, heat flux, and total heat flow based on internal heat generation, conduction, convection, contact resistance and radiation conditions. Thermal orthotropic materials are available.

Transient Thermal: The time dependent thermal study is defined by $M\dot{T} + KT(t) = Q(t)$ where M is the specific heat matrix (also called the thermal mass matrix), K is the conduction and convection matrix, and Q is the combined nodal heat flow vector due to internal sources, convection, radiation, and given heat flux. Given the initial conditions, and boundary conditions the system is time integrated for the model temperatures, $T(t)$.

The SW Simulation software also offers several types of nonlinear studies. Like most commercial finite element systems SW Simulation has many capabilities and the average user only utilizes a few of them.

Table 2.1.   Selected SW Simulation capabilities.

*	Angular acceleration	Nonlinear analysis
*	Angular velocity	Optimization analysis
*	Assembly analysis	* Orthotropic materials
*	Body loads: gravity and centrifugal	p-adaptive analyses
*	Buckling load factors and modes	* Plot customization
	Connectors (Pin, bolt, etc.)	* Prescribed non-zero restraints
	Contact analysis with friction	* Principal stresses
*	Deformation plot	* Probe and list-by-entity tools
*	Directional pressure and force	Publish eDrawings of results
*	Directional restraints	* Reaction force result plot
*	Displacement plots	Remote mass
*	Dynamic section and iso-plots	* Restrain edges, faces and vertices
*	Edit material library	* Result graphs and listings
	Element result plots	S-N fatigue curves
*	Factor of safety calculation and plot	* Strain and displacement analyses
	Fatigue analysis and plots	* Stress analysis
*	Fixed restraints on faces	* Stress contour plots
*	Force on edges, faces and vertices	Stress error estimate
*	Frequency analyses	* Symmetry restraints
	h-adaptive analyses	* Temperature distribution
*	Heat flux result plots	* Temperature gradient plots
*	Heat sources	Temperature dependent properties
*	Heat transfer analysis	Thermal contact resistance
	Import SW Flow loads	* Thermal stress analysis
	Import SW Motion loads	Thermostat controlled heat generation
	Interference or shrink fit	* Thin parts, sheet metal using shells
*	Large displacement analysis	* von Mises equivalent stress
*	Multibody part analysis	* Weldment analysis using beam elements

Table 2.1 lists those abilities that the author thinks are most useful. Those with an asterisk are illustrated in this book. SW Simulation comes with a very good set of tutorials that serve to illustrate all of the above analysis capabilities.

Here SW Simulation will be introduced by examples intended to show basic capabilities generally not covered in the tutorials.

Also some tutorials focus on which icons to pick, and do not have the space to discuss good engineering practice. Here, one goal is to introduce such concepts based on the author's decades of experience in applying finite element (FE) methods. For example, you should always try to validate a FE calculation with approximate analytic solutions and/or a different type of finite element model. This is true for even experienced persons using a program with which they have not had extensive experience. Sometimes you might just misunderstand the supporting documentation and make a simple input error. In the author's opinion, the book by Adams and Askenazi [1] is one of the best practical overviews of the interaction of modern solid modeling (SM) software and general finite element software (such as SW Simulation), and the many pitfalls that will plague many beginners. It points out that almost all FE studies involve assumptions and approximations and the user of such tools should be conscious of them and address them in any analysis or design report. You are encouraged to read it.

2.2. Element Types and Shapes

SW Simulation currently includes solid continuum elements, curved surface shell elements (thin and thick) and truss and frame line elements. Solid elements have only displacement degrees of freedom. General shells have both displacement and rotational degrees of freedom at each node. However, membrane shells (like plane stress elements) have only displacement DOF. The shells are triangular with three vertex nodes or three vertex and three mid-edge nodes (Figure 2.1). The solids are tetrahedra with four vertex nodes or four vertex and six mid-edge nodes. Solid and membrane shell elements use linear and quadratic interpolation for the solution based on whether they have two or three nodes on an edge. The linear elements are also called simplex elements because their number of vertices is one more than the dimension of the space.

Solid elements have their stresses and strains recovered at a number of tabulated locations inside the elements. Stress or strain results from adjacent solids are averaged at their common nodes. Shells

Fig. 2.1. SW Simulation shell (left) and solid element types.

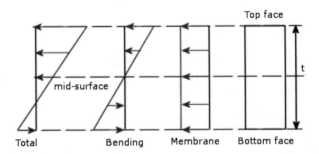

Fig. 2.2. Shells superimpose transverse bending and in-plane stresses.

are approximate solids represented by their mathematical surface geometry and their thickness. Shells have their stresses and strains reported on their "top" (half-thickness above the mid-surface), mid-surface, and "bottom" (half-thickness below the mid-surface) locations, as sketched in Figure 2.2. Generally, the combination of transverse bending stresses and in-plane (membrane) stresses cause the top and bottom of the shell to have stresses of opposite signs.

You should always examine the mesh before starting an analysis run. The size of each element indicates a region where the solution is approximated (piecewise) by a spatial polynomial. Most finite element systems, including SW Simulation, use linear or quadratic polynomials in each element. You can tell by inspection which is being used by looking at an element edge. If that line has two nodes the polynomial is linear. If three nodes, the polynomial is quadratic.

2.3. Element Interpolations

Let $T(x, y, z)$ denote an entity to be interpolated within an element and let x, y, and z be the local element coordinates. You can relate the number of nodes on an element to the number of polynomial coefficients (c_k) in the local element spatial approximation, as outlined below:

Linear element type: Straight edge line, or straight bar–2 nodes,

$$T(x) = c_1 + c_2 x;$$

Straight edged triangle membrane shell, or tetrahedron face–3 nodes,

$$T(x, y) = c_1 + c_2 x + c_3 y;$$

Straight edged, flat faced, tetrahedron–4 nodes,

$$T(x, y, z) = c_1 + c_2 x + c_3 y + c_4 z.$$

Therefore, the solution gradient (first derivatives) in this type of element is constant and many elements are required to get good results. In SW Simulation a mesh of linear elements is called a "Draft Mesh".

Quadratic element type: Edge line–3 nodes, $T(x) = c_1 + c_2 x + c_3 x^2$; Curved triangular membrane shell, or tetrahedron face–6 nodes,

$$T(x, y) = c_1 + c_2 x + c_3 y + c_4 x^2 + c_5 xy + c_6 y^2;$$

General curved tetrahedron–10 nodes,

$$T(x, y, z) = c_1 + c_2 x + c_3 y + c_4 z + c_5 x^2 + c_6 xy$$
$$+ c_7 y^2 + c_8 xz + c_9 yz + c_{10} z^2.$$

These are called complete quadratic elements because there are no terms missing in the quadratic polynomial. Their gradients are complete linear polynomials in three-dimensional space. Therefore, the solution gradient, and strains and heat fluxes, in these elements vary piecewise linearly in space and fewer quadratic elements are required for a good solution. SW Simulation refers to quadratic elements as a "Quality Mesh". The above comments refer to flat shells loaded only in their plane. When flat or curved shells are loaded normal to their surface a more complicated set of interpolations are used to include their transverse bending behavior.

Note that if you set $z = 0$ in the quadratic (10 DOF) solid, to restrict the interpolation to a particular face triangular of the element, you obtain the previous (6 DOF) quadratic triangle. Likewise, if you set both z and y to zero, to restrict the interpolation to a particular edge of the element, you obtain the previous quadratic (3 DOF) line element. Clearly, all of the elements can interpolate data that happen to be constant (like a RBM) with $c_k = 0$, $k \geq 2$. It is less clear, but easy to prove, that all of the quadratic interpolations can reduce to the corresponding (line, surface, volume) linear interpolation if that is the exact solution.

These polynomial interpolations within an element mean that the primary unknown (displacement or temperature) is continuous within the element (has an infinite number of derivatives, C^∞) and across the element interfaces (but only the value is shared with its neighbor, C^0). But, the gradient of the primary unknown is discontinuous across elements, whereas the exact gradient value is continuous in a homogeneous material. The amount of discontinuity between element gradients is reduced as the element size is reduced. For example, the two colored quadratic surfaces in Figure 2.3 could represent the temperature distribution through the two lower adjacent (white) elements. Tangent to the common edge, the temperature and its tangential slope would both be continuous, but the slope normal to that interface is not continuous. Thus, the temperature gradient is discontinuous across each interface, but continuous everywhere inside any single element. Actually, the element interpolations are also used to determine the shape of each element by interpolating between the global position vectors of their nodes or a quadratic shell, the SolidWorks sends the physical

Fig. 2.3.   Exploded view of two quadratic faces.

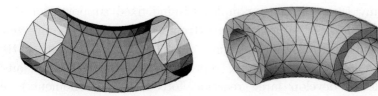

Fig. 2.4.   Piecewise quadratic surface and solid elements.

x, y, z coordinates of each the six nodes to SW Simulation for
building the shell geometry, as seen on the left of Figure 2.4.
Similarly, the element nodes on all surfaces of a solid are defined
by SolidWorks and then the SW Simulation mesh generator builds
the interior tetrahedrons by working in from the bounding surfaces.
While the edges seen in Figure 2.4 would be defined exactly in
SolidWorks as circles, and they "look like" circles in the finite
element mesh they are actually piecewise approximations of a circle.
In other words, the quadratic edge of an element is a segment of
a parabola passing through the three edge nodes that is used to
approximate a segment of a circular arc through the same three
nodes.

2.4. Common Modeling Errors

As noted above, FE models often have small geometric errors. They
can be reduced by mesh refinement and are usually much less
important than other sources of error. Probably the most common
source of error is in selecting the restraint approximations to be
applied to a model. Usually a restraint is applied to a region where
surrounding material has been removed and it is necessary to replace
the missing material with a restraint. Keep in mind that the removed
material must be capable of supplying the assumed restraint or you
may introduce a very large error. Sometimes you should move the
restraints further away from the part of main interest by including
a small region of the supporting material to which you apply the
restraints.

Many tutorials and examples assume fixed supports for simplicity. True fixed supports are extremely rare. They require zero movement of the support region. That in turn means that the removed supporting material (represented by the restraint) must be able to develop large reaction forces (and/or moments). A fixed support assumes the material can convey both tension and compression reaction forces locally as needed. Yet some supports can only convey tension while others can only resist compression. Fixed support assumptions tend to under estimate the stresses in the part of interest, but over estimate the resisting stresses (reactions) in the removed material replaced by our simplified engineering assumptions (the restraint type).

Loadings are also not as clear elementary examples suggest. Is a force applied as a point load, a line load, a surface load, etc.? That is, where and how a load is applied is usually an assumption. Likewise, the magnitude of a force or other load may be a reasonable guess or it may be given by established design codes. In thermal studies the convection coefficients vary over a wide range and you may have to run different studies with the high and low values.

The nature of the equations being solved is such that the computed reactions are essentially always equal and opposite to the resultant actual applied loads, not necessarily the loads you though that you applied. Reaction data are available in SW Simulation and you should always check them.

Common "standardized" materials have mechanical and thermal properties that are relatively well known and are built into the SW Simulation materials library. However, even those materials have some range in their values that are not represented in a single number stored in a table. Many important properties, like the modulus of elasticity, are experimentally measured to only two or three significant figures. Yet a materials table frequently gives average values or values converted from other units to a misleading six or seven significant figures. So usually the computed displacements are only accurate to three or four digits and the stresses to two or three digits.

2.5. Infinite Corner Gradients

It has already been mentioned that while point forces (or heat sources) are commonly used in an FEA, they cause theoretically infinite gradients (stresses or heat flux) at such points. A related consideration is the treatment of reentrant corners or edges in the part. All elliptical differential equations (stress analysis, heat transfer, potential flow, electrostatics, etc.) will have a local singularity (infinite radial derivative) where the boundary of the domain contains a reentrant corner and/or a reentrant edge. If you fillet any such corner or edge then the singularity does not occur. However, if you defeature the fillet and return to sharp corners then you introduce these singular derivatives. The interior angle, A, between the corner faces is the governing feature. For values less than or equal to 180 degrees there is no reentrant geometry or local singularity. Otherwise, a singularity occurs as the radius, r, goes to zero in the corner.

For a scalar unknown like potential flow, or heat transfer, the local solution around the corner varies with the local angle, a, as: $\varphi(r, a) = K r^{\pi/A} f(a)$. Here K is the intensity (importance) of the singularity and $\beta = \pi/A$ is the strength of the singularity. A right angle corner ($A = 3\pi/2$) is relatively weak ($\beta = 2/3$) while a slit or crack ($A = 2\pi$) is the strongest with $\beta = 1/2$. The radial gradient, in polar coordinates, will go to infinite any time the angle is more than 180 degrees ($\beta < 1$): $\partial \varphi / \partial r = (K\pi/A) \, r^{(\beta-1)} f(a)$. For a crack the radial gradient is proportional to one over the square root of the radius: $\partial \varphi / \partial r \propto 1/r^{1/2}$.

Therefore, the radial gradient goes to infinite as the radius r goes to zero at the corner. Since the local error in a FE solution is proportional to the product of the gradient and the element size, in theory you need to have very small element sizes (almost zero) at such reentrant geometries. In practice, most corners are not mathematically sharp and some small radius develops during manufacture. Still, the gradients can be very large. When the solid part has a mathematically sharp corner, the false infinite stresses (or strains, or heat flux) develops at the corner point. When you contour

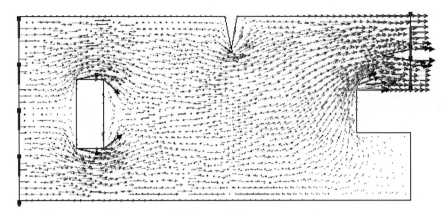

Fig. 2.5.   Gradient vectors must change as they pass around corners.

singular results the false high values may lead you to overlook true high gradient values at other locations. At times you will need to reduce the automatically selected maximum contour level in your plots.

The actual intensity (importance), K, of the corner singularity depends on how the far field solution is distributed relative to the centerline of the corner. Corners with the same angle will have different importance depending on whether the far field solution is changing parallel-to or perpendicular-to the corner centerline. This is illustrated graphically in Figure 2.5 where there are six 90 degree corners and two with larger interior angles. In this case, the default plot refers to heat flux vectors, but the same Poisson equation would apply to ideal fluid flow where the primary unknown is the velocity potential and the gradient vectors would be the actual fluid velocity vectors (listed as the heat flux vectors in the last chapter). As can be seen from Figure 2.5, reentrant corner singularities usually affect the solution gradients, but their importance depends on their location in the part. Apply reasonable mesh refinements near all such corner regions.

3

Concepts of Stress Analysis

3.1. Introduction

Here the concepts of stress analysis will be stated in a finite element context. That means that the primary unknown will be the (generalized) displacements. All other items of interest will mainly depend on the displacement gradient and therefore will be less accurate than the displacements. Stress analysis covers several common special cases to be mentioned later. Here only two formulations will be considered initially. They are the solid continuum form and the shell form. Both are offered in SW Simulation. They differ in that the continuum form utilizes only displacement vectors, while the shell form utilizes displacement vectors *and* infinitesimal rotation vectors at the element nodes.

As illustrated in Figure 3.1, the solid elements have three translational degrees of freedom (DOF) as nodal unknowns, for a total of 12 or 30 DOF for the linear and quadratic solids, respectively. The shell elements have three translational degrees of freedom as well as three rotational degrees of freedom, for a total of 18 or 36 DOF. The difference in DOF types means that moments or couples can only be applied directly to shell models. Solid elements require that couples be indirectly applied by specifying a pair of equivalent pressure distributions, or an equivalent pair of equal and opposite forces at two nodes on the body.

Stress transfer takes place within, and on, the boundaries of a solid body. The **displacement vector**, u, at any point in the

Fig. 3.1. Degrees of freedom for frames and shells; solids and trusses.

continuum body has the units of meters [m], and its components are the primary unknowns. The components of displacement are usually called u, v, and w in the x-, y-, and z-directions, respectively. Therefore, they imply the existence of each other, $\boldsymbol{u} \leftrightarrow (u, v, w)$. All the displacement components vary over space. As in the heat transfer case (covered later), the gradients of those components are needed but only as an intermediate quantity. The displacement gradients have the units of [m/m], or are considered dimensionless. Unlike the heat transfer case where the gradient is used directly, in stress analysis the multiple components of the displacement gradients are combined into alternate forms called strains.

In 1D, the normal strain is just the ratio of the change in length over the original length, $\varepsilon_x = \partial u / \partial x$. In 2D and 3D, both normal strains and shear strains exist. The normal strains involve only the part of the gradient terms parallel to the displacement component. In 2D they are $\varepsilon_x = \partial u / \partial x$ and $\varepsilon_y = \partial v / \partial y$. They would cause a change in volume, but not a change in shape of the rectangular differential element. A shear strain causes a change in shape. The total angle change (from 90 degrees) is used as the engineering definition of the shear strain. The shear strains involve a combination of the components of the gradient that are perpendicular to the displacement component. In 2D, the engineering shear strain is $\gamma = (\partial u / \partial y + \partial v / \partial x)$. Strain has one component in 1D, three components in 2D, and six components in 3D. The 2D strains are commonly written as a column vector: $\boldsymbol{\varepsilon} = (\varepsilon_x \quad \varepsilon_y \quad \gamma)^T$.

Stress is a measure of the force per unit area acting on a plane passing through the point of interest in a body. The above geometrical data (the strains) will be multiplied by material properties to define a new physical quantity, the **stress**, which is in initially

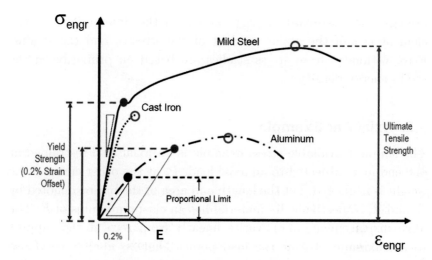

Fig. 3.2.   Hooke's Law for initial linear stress–strain, $\sigma = E\varepsilon$.

directly proportional to the strains. This is known as **Hooke's Law**: $\sigma = E\varepsilon$, (see Figure 3.2) where the square **material matrix, E,** contains the elastic modulus, and Poisson's ratio of the material. The 2D stresses are written as a corresponding column vector, $\sigma = (\sigma_x \quad \sigma_y \quad \tau)^T$. Unless stated otherwise, the applications illustrated here are assume to be in the linear range of a material property.

The 2D and 3D stress components are shown in Figure 3.3. The normal and shear stresses represent the normal force per unit area and the tangential forces per unit area, respectively. They have the units of $[N/m^2]$, or $[Pa]$, but are usually given in $[MPa]$. The **strain**

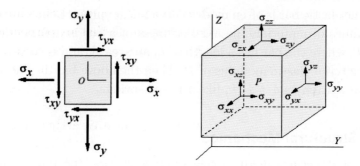

Fig. 3.3.   Stress components in 2D (left, plane stress) and 3D solids.

energy (or **potential energy**) stored in the differential material element is half the scalar product of the stresses and the strains. Error estimates from stress studies are based on primarily on the strain energy density.

3.2. Axial Bar Example

The simplest available stress example is an axial bar, restrained at one end and subjected to an axial load, P, at the other end and the weight is neglected. Let the length and area of the bar be denoted by L, and A, respectively. Its material has an elastic modulus of E. The axial displacement, $u(x)$, varies linearly from zero at the support to a maximum of δ at the load point. That is, $u(x) = x\delta/L$, so the axial strain is $\varepsilon_x = \partial u/\partial x = \delta/L$, which is a constant. Likewise, the axial stress is everywhere constant, $\sigma = E\varepsilon = E\delta/L$ which in the case simply reduces to $\sigma = P/A$. Like many other more complicated problems, the stress here does not depend on the material properties, but the displacement always does, $\delta = PL/EA$. You should always carefully check both the deflections and stresses when validating a finite element solution.

Since the assumed displacement is linear here, any finite element model would give exact deflection and the constant stress results. However, if the load had been the distributed bar weight the exact displacement would be quadratic in x and the stress would be linear in x. A quadratic element mesh would give exact stresses and displacements everywhere.

The elastic bar is often modeled as a linear spring. In introductory mechanics of materials the axial stiffness of a bar is defined as $k = EA/L$, where the bar has a length of L, an area A, and is constructed of a material elastic modulus of E. Then the above bar displacement can be written as $\delta = P/k$, like a linear spring.

3.3. Structural Mechanics

Modern structural analysis relies extensively on the finite element method. The most popular integral formulation, based on the variational calculus of Euler, is the Principle of Minimum Total Potential Energy. Basically, it states that the displacement field that satisfies the essential displacement boundary conditions and minimizes the total potential energy is the one that corresponds to the state of static equilibrium. This implies that displacements are our primary unknowns. They will be interpolated in space as will their derivatives, and the strains. The total potential energy, Π, is the strain energy, U, of the structure minus the mechanical work, W, done by the external forces. From introductory mechanics, the mechanical work, W, done by a force is the scalar dot product of the force vector, $\boldsymbol{F}$, and the displacement vector, $\boldsymbol{u}$, at its point of application.

The well-known linear elastic spring will be reviewed to illustrate the concept of obtaining equilibrium equations from an energy formulation. Consider a linear spring, of stiffness k, that has an applied force, F, at the free (right) end, and is restrained from displacement at the other (left) end, as in Figure 3.4. The free end undergoes a displacement of Δ. The work done by the single force is $W = \vec{\Delta} \circ \vec{F} = \Delta_x F_x = uF$. The spring stores potential energy due to its deformation (change in length). Here we call that strain energy. That stored energy is given by $U = (1/2)k\Delta_x^2$. Therefore, the total potential energy for the loaded spring is

$$\Pi = \frac{1}{2}k\Delta_x^2 - \Delta_x F_x.$$

The equation of equilibrium is obtained by minimizing this total potential energy with respect to the unknown displacement, Δ_x.

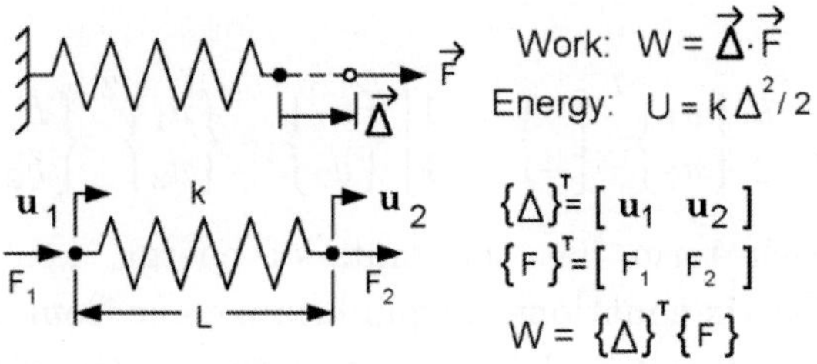

Fig. 3.4.   Classic (top) and general linear spring element.

That is,

$$\frac{\partial \Pi}{\partial \Delta_x} = 0 = \frac{2}{2}k\Delta_x - F_x.$$

This simplifies to the common single scalar equation $k\Delta_x = F$, which is the well-known equilibrium equation for a linear spring. This example was slightly simplified, since we started with the condition that the left end of the spring had no displacement (an essential or Dirichlet boundary condition). Next we will consider a spring where either end can be fixed or free to move. This will require that you both minimize the total potential energy and impose the given displacement restraint.

Now the spring model has two end displacements, u_1 and u_2, and two associated axial forces, F_1 and F_2. The net deformation of the bar is $\delta = u_2 - u_1$. Denote the total vector of displacement components as $\vec{\Delta} = \{u\} = \{{u_1 \atop u_2}\}$ and the associated vector of forces as $\vec{F} = \{F\} = \{{F_1 \atop F_2}\}$. The mechanical work done on the spring is $W = \{u\}^T\{F\} = u_1F_1 + u_2F_2$. Then the spring's strain energy is

$$U = \frac{1}{2}\{u\}^T[k]\{u\} = \frac{1}{2}k\delta^2,$$

where the "spring stiffness matrix" is found to be

$$[k] = k\begin{bmatrix} 1 & -1 \\ -1 & 1 \end{bmatrix}.$$

The total potential energy, Π, becomes

$$\Pi = \frac{1}{2}\{u\}^T[k]\{u\} - \{u\}^T\{F\}$$

or

$$\Pi = \frac{k}{2}\begin{Bmatrix} u_1 \\ u_2 \end{Bmatrix}^T \begin{bmatrix} 1 & -1 \\ -1 & 1 \end{bmatrix}\begin{Bmatrix} u_1 \\ u_2 \end{Bmatrix} - \begin{Bmatrix} u_1 \\ u_2 \end{Bmatrix}^T\begin{Bmatrix} F_1 \\ F_2 \end{Bmatrix}.$$

Note that each term has the units of energy, i.e., force times length. The matrix equations of equilibrium come from satisfying the displacement restraint and the minimization of the total potential energy with respect to each and every displacement component.

The minimization requires that the partial derivative of all the displacements vanish: $\partial\Pi/\partial\{u\} = \{0\}$, or $\partial\Pi/\partial u_j = 0_j$. That represents the first stage system of algebraic equations of equilibrium for the elastic system:

$$k \begin{bmatrix} 1 & -1 \\ -1 & 1 \end{bmatrix} \begin{Bmatrix} u_1 \\ u_2 \end{Bmatrix} = \begin{Bmatrix} F_1 \\ F_2 \end{Bmatrix}.$$

These two symmetric equations do not yet reflect the presence of any essential boundary condition on the displacements. Therefore, no unique solution exists for the two displacements due to applied forces (the axial RBM has not been eliminated). Mathematically, this is clear because the square matrix has a zero determinant and cannot be inverted. If all of the displacements are known, you can find the applied forces. For example, if you had a rigid body translation of $u_1 = u_2 = C$ where C is an arbitrary constant you clearly get $F_1 = F_2 = 0$. If you stretch the spring by two equal and opposite displacements; $u_1 = -C$, $u_2 = C$ and the first row of the matrix equations gives $F_1 = -2kC$. The second row gives $F_2 = 2kC$, which is equal and opposite to F_1, as expected. Usually, you know some of the displacements and some of the forces. Then you have to manipulate the matrix equilibrium system to put it in the form of a standard linear algebraic system where a known square matrix multiplied by a vector of unknowns is equal to a known vector: $[A]\{x\} = \{b\}$.

3.4. Equilibrium of a Single Restrained Element

In basic physics, we learn to utilize springs connected in series and/or in parallel. Elastic bodies behave as springs. For 1D bars, shafts, and beams we can utilize their analytic stiffness relations to derive approximate (and sometimes exact) analytic solutions at discrete points on the body. That allows us to treat axial forces, axial torsion, and transverse forces in simple models.

As noted above, the equilibrium relation for a simple spring between its stiffness, k, displacement, u, and axial force, f, is usually seen as: $ku = f$, or $u = f/k$. This simple form arises only because we initially assumed that one end of the spring was restrained from displacement (immovable), and the opposite end was subjected to a constant force. If the spring model is generalized to allow either end to be restrained or loaded then the equilibrium equation takes a matrix form:

$$k \begin{bmatrix} 1 & -1 \\ -1 & 1 \end{bmatrix} \begin{Bmatrix} u_1 \\ u_2 \end{Bmatrix} = \begin{Bmatrix} f_1 \\ f_2 \end{Bmatrix},$$

where the subscripts 1 and 2 refer to the left and right ends of the spring, respectively. Note that the determinant of the matrix is zero. That is because were must later add restraint information about at least one end to obtain a unique physical solution. For example, assume that the left node has a known displacement (which is usually zero) and the right end has a known force, $f_2 = F$. The unknowns are the right displacement, u_2, and the left end reaction force, say $f_1 = R$. The revised analytic equilibrium relation is

$$k \begin{bmatrix} 1 & -1 \\ -1 & 1 \end{bmatrix} \begin{Bmatrix} u_{given} \\ u_2 \end{Bmatrix} = \begin{Bmatrix} R \\ F \end{Bmatrix},$$

and the independent displacement is found from the second row:

$$k[-u_{given} \quad u_2] = F,$$

$$u_2 = u_{given} + \frac{F}{k}.$$

This is the same as the common form when u_{given} is zero, namely $u_2 = F/k$. Now the reaction force necessary to maintain u_{given} is obtained from the first row of the matrix system:

$$k \left[u_{given} - \left(u_{given} + \frac{F}{k} \right) \right] = R$$

or simply $R = -F$, as expected.

An elastic bar acts like a simple spring. However, in addition to end point loads it can have distributed mechanical loads per unit length, and/or thermal loading due to a temperature change, say ΔT, over its length. The resultants of such effects are lumped at the ends as additional point loads. For a linear bar with a cross-sectional area, A, length, L, a distributed load per unit length of w_1 on the left and w_2 on the right, and material with an elastic modulus of E and a coefficient of thermal expansion of α, the corresponding matrices are

$$\frac{EA}{L}\begin{bmatrix} 1 & -1 \\ -1 & 1 \end{bmatrix}\begin{Bmatrix} u_1 \\ u_2 \end{Bmatrix} = \begin{Bmatrix} f_1 \\ f_2 \end{Bmatrix} + \frac{L}{6}\begin{bmatrix} 2 & 1 \\ 1 & 2 \end{bmatrix}\begin{Bmatrix} w_1 \\ w_2 \end{Bmatrix} + \alpha \Delta T E A\begin{Bmatrix} -1 \\ 1 \end{Bmatrix},$$

where, again, the f_k terms represent external point loads or reactions. The ratio $k = EA/L$ is called the axial stiffness of a bar. If the line load is constant then the second load vector (transposed) reduces to $F_w^T = wL[1\ 1]/2$ which places half the total applied line load at each end of the bar. The displacement between the two ends was assumed to be linear. That causes the strain to be constant, $\varepsilon = \Delta L/L = (u_2 - u_1)/L$, which is not correct for a non-zero line load, $w(x)$.

A minor change in interpretation yields a similar relation for the torsion of a straight shaft. The displacements convert to the angle of twist about the axis, the distributed load becomes a torque per unit length $t(x)$, and point forces convert to torque couples, T, and temperature effects do not appear. The elastic modulus is replaced by the shear modulus, G, and the area is replaced by the polar moment of inertia of the section, J, to define the torsional stiffness of a shaft, $k = GJ/L$. The torsional equilibrium relations are:

$$\frac{GJ}{L}\begin{bmatrix} 1 & -1 \\ -1 & 1 \end{bmatrix}\begin{Bmatrix} \theta_1 \\ \theta_2 \end{Bmatrix} = \begin{Bmatrix} T_1 \\ T_2 \end{Bmatrix} + \frac{L}{6}\begin{bmatrix} 2 & 1 \\ 1 & 2 \end{bmatrix}\begin{Bmatrix} t_1 \\ t_2 \end{Bmatrix}.$$

The accuracy and complexity of a bar model (or a shaft model) is increased by basing it on quadratic interpolation, based on three nodes per element. For the usual case where the second node is at the midpoint, the matrix equilibrium equations for a single bar element becomes:

$$\frac{EA}{3L}\begin{bmatrix} 7 & -8 & 1 \\ -8 & 16 & -8 \\ 1 & -8 & 7 \end{bmatrix}\begin{Bmatrix} u_1 \\ u_2 \\ u_3 \end{Bmatrix}$$

$$= \begin{Bmatrix} f_1 \\ f_2 \\ f_3 \end{Bmatrix} + \frac{L}{30}\begin{bmatrix} 4 & 2 & -1 \\ 2 & 16 & 2 \\ -1 & 2 & 4 \end{bmatrix}\begin{Bmatrix} w_1 \\ w_2 \\ w_3 \end{Bmatrix} + \alpha \Delta T E A\begin{Bmatrix} -1 \\ 0 \\ 1 \end{Bmatrix}$$

for a quadratic (three point) line load. A constant line load resultant vector is

$$\boldsymbol{F}_w^T = \frac{wL}{6}[1 \quad 4 \quad 1], \quad (\text{for } w_1 = w_2 = w_3 = w)$$

while an increasing triangular load yields

$$\boldsymbol{F}_w^T = \frac{wL}{3}[0 \quad 2 \quad 1], \quad \left(\text{for } w_1 = 0, w_2 = \frac{w}{2}, w_3 = w\right).$$

Generally, you want to substitute the three nodal line load values, w_k, to obtain the resultant vector, $\boldsymbol{F}_w$, before assembling with other elements. If you want to interpolate the approximate displacement solution or gradient between the nodes, the interpolation functions are (for non-dimensional position $r = x/L$):

$$u(x) = u_1(1 - 3r + 2r^2) + u_2(4r - 4r^2) + u_3(-r + 2r^2),$$

$$\varepsilon(x) = \frac{du}{dx} = \frac{u_1(-3 + 4r) + u_2(4 - 8r) + u_3(-1 + 4r)}{L}.$$

At node 1 where $r = 0$ the displacement and strain become: $u(0) = u_1$, and $\varepsilon(0) = (4u_2 - u_3 - 3u_1)$. For a rigid body translation, $u_1 = u_2 = u_3 = c$, by definition. Then $\varepsilon(0) = 0$ and there is no strain or stress at that point. Actually, the strain everywhere, $\varepsilon(x)$, vanishes for that special case.

3.5. General Equilibrium Matrix Partitions

The above small example gives insight to the most general form of the algebraic system resulting from only minimizing the total potential energy: a singular matrix system with more unknowns than equations. That is because there is not a unique equilibrium solution to the problem until you also apply the essential (Dirichlet) boundary conditions on the displacements. The algebraic system can be written in a general partitioned matrix form that more clearly defines what must be done to reduce the system by utilizing essential boundary conditions.

For an elastic system of any size, the full, symmetric matrix equations obtained by minimizing the energy can always be rearranged into the following partitioned matrix form:

$$\begin{bmatrix} K_{uu} & K_{ug} \\ K_{gu} & K_{gg} \end{bmatrix} \begin{Bmatrix} \Delta_u \\ \Delta_g \end{Bmatrix} = \begin{Bmatrix} F_g \\ F_u \end{Bmatrix},$$

where Δ_u represents the unknown nodal displacements, and Δ_g represents the given essential boundary values (restraints, or fixtures) of the other displacements. The stiffness sub-matrices K_{uu} and K_{gg} are square, whereas K_{ug} and K_{gu} are rectangular matrices. In a finite element formulation all of the coefficients in the K matrices are known. The resultant applied nodal loads are in sub-vector F_g and the F_u terms represent the unknown generalized reactions forces associated with essential boundary conditions. This means that after the enforcement of the essential boundary conditions the actual remaining unknowns are Δu and F_u. Only then does the net number of unknowns correspond to the number of equations. But, they must be re-arranged before all the remaining unknowns can be computed.

Here, for simplicity, it has been assumed that the equations have been numbered in a manner that places rows associated with the given displacements (essential boundary conditions) at the end of the system equations. The above matrix relations can be rewritten as two sets of matrix identities:

$$K_{uu}\Delta_u + K_{ug}\Delta_g = F_g,$$
$$K_{gu}\Delta_u + K_{gg}\Delta_g = F_u.$$

The first identity can be solved for the unknown displacements, Δ_u, by putting it in the standard linear equation form by moving the known product $K_{ug}\Delta_g$ to the right side. Most books on numerical analysis assume that you have reduced the system to this smaller, non-singular form (K_{uu}) before trying to solve the system. Inverting the smaller non-singular square matrix yields the unique equilibrium displacement field:

$$\Delta_u = K_{uu}^{-1}(F_g - K_{ug}\Delta_g).$$

The reaction forces can then be recovered, if desired, from the second identity:

$$F_u = K_{gu}\Delta_u + K_{gg}\Delta_g.$$

In most applications, these reaction data have physical meanings that are important in their own right, or useful in validating the solution. However, this part of the calculation is optional.

3.6. Assembly of Multiply Element Connections

When a system of springs is combined in series or parallel, they share (scatter or sum) square matrix diagonal terms and source vector row terms at any connecting points (or rows and columns) in the system matrix equilibrium equation, see Figure 3.5. The size of the overlapping summed (scattered) regions depends on how many degrees of freedom are at a shared connection point. Often it is only a single shared term, but for a beam there is usually a deflection and slope at the connection and two rows are summed. The sum (assembly) of the nodal forces from the individual spring internal forces (or sources) at a node connection equals the resultant external point force there. Often the resultant is zero (the external force is absent). If the node point DOF is restrained then the external resultant is an unknown point reaction.

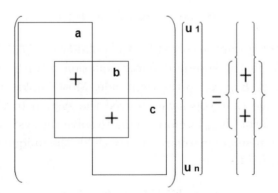

Fig. 3.5.   Overlapping matrix sums (scatters) at shared connections.

The finite element analytical forms of the element stiffness relations and the assembled (scattered) equations of equilibrium can prove, at times, to be useful in obtaining closed form approximations of a particular physical problem. Often that problem is a simplified model of a real problem that will be used to validate or check the large numerical calculation.

Consider two springs, say a and b, connected in series. Restrain let the first (left) point while the third (right) point has a known force, F. The second (middle) has no resultant external force applied. Then the assembled 3 by 3 system equilibrium matrix form become:

$$\begin{bmatrix} k^a & -k^a 2 & 0 \\ -k^a & (k^a + k^b) & -k^b \\ 0 & -k^b & k^b \end{bmatrix} \begin{Bmatrix} u_1 = u_{given} \\ u_2 \\ u_3 \end{Bmatrix} = \begin{Bmatrix} f_1^a \\ f_2^a + f_1^b \\ f_2^b \end{Bmatrix} = \begin{Bmatrix} R \\ 0 \\ F \end{Bmatrix}.$$

This is a set of three equations for three unknowns: u_2, u_3 and R. But, they are singular and cannot be solved until it is modified for all known displacements. Likewise, the reaction, R, cannot be found before the displacements are known, so we use the last two rows to do that. Enforce the essential boundary condition by multiplying the first column of the system by the known value of u_1, and move that column to the right hand side (with u_{given} zero):

$$\begin{bmatrix} (k^a + k^b) & -k^b \\ -k^b & k^b \end{bmatrix} \begin{Bmatrix} u_2 \\ u_3 \end{Bmatrix} = \begin{Bmatrix} 0 \\ F \end{Bmatrix} - u_{given} \begin{Bmatrix} -k^a \\ 0 \end{Bmatrix} = \begin{Bmatrix} 0 \\ F \end{Bmatrix}$$

so inverting the reduced 2 by 2 square equilibrium matrix (which is now non-singular) gives the unknown displacement values:

$$\begin{Bmatrix} u_2 \\ u_3 \end{Bmatrix} = \frac{1}{d} \begin{bmatrix} k^b & k^b \\ k^b & (k^a + k^b) \end{bmatrix} \begin{Bmatrix} 0 \\ F \end{Bmatrix} = \frac{F}{d} \begin{Bmatrix} k^b \\ k^a + k^b \end{Bmatrix},$$

where the determinant of the reduced square matrix is $d = k^a k^b$. Returning to the first row in the original system of three equations gives the reaction necessary to keep u_1 at zero:

$$\frac{F}{k^a k^b} \begin{bmatrix} 0 & -k^a k^b & 0 \end{bmatrix} = -F = R.$$

That reaction is equal and opposite to the applied force, as expected. This concludes the first analytic matrix solution example.

As a second analytic example, let the above springs represent a linear displacement bar with the first half having an area of $2A$ while the last half has an area of A. The two axial stiffness's $k = EA/L$ are $k^a = E(2A)/(L/2) = 4EA/L$ and $k^b = EA/(L/2) = 2EA/L$. Then the deflections are

$$\begin{Bmatrix} u_2 \\ u_3 \end{Bmatrix} = \frac{FL}{4EA} \begin{Bmatrix} 2 \\ 3 \end{Bmatrix}.$$

So you can see that the smaller right end element's relative deflection, $(u_3 - u_2)$, is only half as much as the displacement of the larger element on the left.

3.7. Solution Phases for Compound Elastic Bars

Because there was no distributed load in the above example, it is simple enough to see that the force in each segment of the bar is clearly just F. However, there is a standard process for finding the end reactions on each spring from the known displacements of the system. Simply gather the known displacements for each individual bar and substitute them into an element's individual matrix equilibrium equation (including it's line load vector). To illustrate all these features, and to explain the stress recovery, a numerical example will be given for a compound axial bar. A large compound hanging bar made of an upper steel section and lower brass section carries its distributed weight (gravity load) and a point end load of 10,000 lb (see Figure 3.6). There is no thermal loading. The tabulated properties of the system are:

El	Length	Area	Modulus	Specific weight	Connections	
1	420″	10 sq.in.	30e6 psi	0.283 lb/cubic in.	1	2
2	240″	8 sq.in.	13e6 psi	0.300 lb/cubic in.	2	3

Those data show that the axial stiffness of the steel bar is $k^a = 7.143e5$ lb/in, its total weight is $W^a = 1{,}188.6$ lb corresponding to a distributed load of $w^a = 2.83$ lb/in. For the lower brass

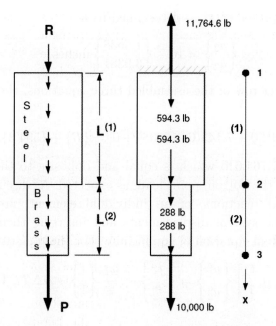

Fig. 3.6.   Steel–brass compound bar with gravity and end forces.

section $k^b = 4.333e5\,\text{lb/in}$, $W^b = 576\,\text{lb}$, $w^b = 2.4\,\text{lb/in}$. Scattering the two axial members gives the assembled system equilibrium equation of:

$$10^5 \begin{bmatrix} 7.143 & -7.143 & 0 \\ -7.143 & (7.143 + 4.333) & -4.333 \\ 0 & -4.333 & 4.333 \end{bmatrix} \begin{Bmatrix} u_1 = 0 \\ u_2 \\ u_3 \end{Bmatrix}$$

$$= \begin{Bmatrix} R \\ 0 \\ 10{,}000 \end{Bmatrix} + \frac{1}{2} \begin{Bmatrix} 1{,}188.6 \\ 1{,}188.6 + 576 \\ 576 \end{Bmatrix},$$

where R is the unknown top reaction force. After modifying the system equations for the essential boundary condition, the bottom two rows become

$$10^5 \begin{bmatrix} 11.476 & -4.333 \\ -4.333 & 4.333 \end{bmatrix} \begin{Bmatrix} u_2 \\ u_3 \end{Bmatrix} = \begin{Bmatrix} 882.3 \\ 10{,}288 \end{Bmatrix}.$$

And the displacements are determined to be

$$\begin{Bmatrix} u_2 \\ u_3 \end{Bmatrix} = 10^{-2} \begin{Bmatrix} 1.5638 \\ 3.9381 \end{Bmatrix} \text{ inches.}$$

From the first row of the assembled three equations, the reaction is recovered as

$$10^5[0 - 7.143(0.015638) + 0] = \{R\} + \{594.3\},$$

or $R = -11{,}764.6$ lb which is equal and opposite to the combined weights and the bottom point load, as expected. As discussed above, the two end reactions, of an individual element, are found by gathering the system displacements and inserting them into each bar's individual equation of equilibrium. For the top steel bar:

$$\frac{EA}{L} \begin{bmatrix} 1 & -1 \\ -1 & 1 \end{bmatrix} \begin{Bmatrix} u_1^a \\ u_2^a \end{Bmatrix} = \begin{Bmatrix} f_1^a \\ f_2^a \end{Bmatrix} + \frac{wL}{2} \begin{Bmatrix} 1 \\ 1 \end{Bmatrix} + \alpha \Delta TEA \begin{Bmatrix} -1 \\ 1 \end{Bmatrix},$$

or

$$7.143x10^5 \begin{bmatrix} 1 & -1 \\ -1 & 1 \end{bmatrix} 10^{-2} \begin{Bmatrix} 0 \\ 1.5638 \end{Bmatrix} = \begin{Bmatrix} f_1^a \\ f_2^a \end{Bmatrix} + \frac{1{,}188.6}{2} \begin{Bmatrix} 1 \\ 1 \end{Bmatrix} + \begin{Bmatrix} 0 \\ 0 \end{Bmatrix}$$

$$\begin{Bmatrix} f_1^a \\ f_2^a \end{Bmatrix} = \begin{Bmatrix} -11{,}764.4 \\ 10{,}576 \end{Bmatrix} \text{ lb.}$$

Likewise, for the lower bar the two end reaction forces are

$$\begin{Bmatrix} f_1^b \\ f_2^b \end{Bmatrix} = \begin{Bmatrix} -10{,}576 \\ 10{,}000 \end{Bmatrix} \text{ lb.}$$

These member and system reactions are sketched in Figure 3.7 to illustrate that they are in equilibrium with each other, and with the externally applied loads.

There are consistent finite element methods for estimating stresses in two node analytic springs. However, they are not generally accurate when line loads are present. Instead, the use of basic mechanics of materials and the above member reactions will yield exact stresses at both ends of the member. In the above axial bar example remember that the axial stress, σ, is simply $\sigma = F/A$ where

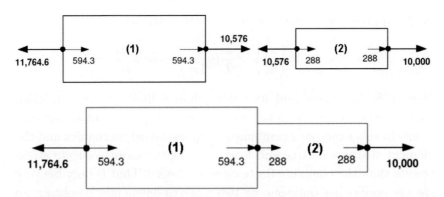

Fig. 3.7. Exploded members, and combined system reaction forces.

F is the axial force acting over area A. For the bars the two end stresses are simply

$$\begin{Bmatrix} \sigma_1^a \\ \sigma_2^a \end{Bmatrix} = \begin{Bmatrix} 11,764.4 \\ 10,576 \end{Bmatrix} \text{lb}/(10\,\text{sq.in}) = \begin{Bmatrix} 1.18 \\ 1.06 \end{Bmatrix} \text{ksi, tension.}$$

$$\begin{Bmatrix} \sigma_1^b \\ \sigma_2^b \end{Bmatrix} = \begin{Bmatrix} 10,576 \\ 10,000 \end{Bmatrix} \text{lb}/(8\,\text{sq.in}) = \begin{Bmatrix} 1.32 \\ 1.25 \end{Bmatrix} \text{ksi, tension.}$$

Based on mechanics of materials (not FEA), assume a linear axial stress distribution between the two ends results in the sketch of stresses shown in Figure 3.8. Note that the system stress is discontinuous due to the change in the two cross-sectional areas (even though the axial force and the displacement are continuous). In this example, the linear stress change in both bars is exact. The consistent finite element theory for the stress in such a bar is only accurate at the mid-point of the bar. It is given by the product of the gradient

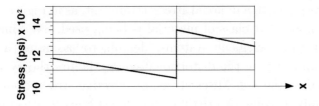

Fig. 3.8. Axial stress distribution in the steel (left) and brass bars.

of the approximate displacements and Hooke's law:

$$\sigma = \frac{E}{L}[-1 \quad 1]\begin{Bmatrix} u_1 \\ u_2 \end{Bmatrix} = \frac{E}{L}(u_2 - u_1) = \frac{E\Delta L}{L} = E\varepsilon.$$

Here, ε is the strain and its value (change in length over original length) is exact only for a bar with a constant point load at one end. Thus, in this example a combination of basis solid mechanics and the element stiffness matrix and load vector gives more accurate stress results than the complete finite element theory. That is only because we are employing only one or two analytic elements, to obtain an analytic answer, whereas full numerical solutions can use hundreds of small elements to give accurate numerical results everywhere in the system. Single bars with more than two nodes each are available for more accurate element displacements and stresses. For example, if you used the quadratic (three-node) elastic bar element given earlier the displacements and stresses would be exact everywhere (for one-dimensional theory). However, the assembled equilibrium equations would be 5 by 5 in size. After enforcing the essential boundary condition you would be left with four simultaneous equations to solve. That is unpleasant to do by hand. Similar analytic matrix solutions will be given later in the chapter on space frames.

3.8. Structural Component Failure

Structural components can be determined to fail by various modes determined by buckling, deflection, natural frequency, strain, or stress. Strain or stress failure criteria are different depending on whether they are considered as brittle or ductile materials. The difference between brittle and ductile material behaviors is determined by their response to a uniaxial stress–strain test, as in Figure 3.2. You need to know what class of material is being used. SW Simulation, and most finite element systems, default to assuming a ductile material and display the distortional energy failure theory which is usually called the von Mises stress, or effective stress, even though it is actually a scalar. A brittle material requires the use of a higher factor of safety.

3.9. Overall Factor of Safety

All aspects of a design have some degree of uncertainty, as does how the design will actually be utilized. For all the reasons cited above, you must always employ a Factor of Safety (FOS). Some designers refer to it as the factor of ignorance. Remember that a FOS of unity means that failure is eminent; it does not mean that a part or assembly is safe. In practice you should try to justify $1 < \text{FOS} < 8$. Several consistent approaches for computing a FOS are given in mechanical design books [9]. They should be supplemented with the additional uncertainties that come from an FEA. Many authors suggest that the factor of safety should be computed as the product of terms that are all ≥ 1. There is a factor for the certainty of the restraint location and type; the certainty of the load region, type, and value; a material factor; a cyclic load factor; and an additional factor if failure is likely to result in human injury. Various professional groups and standards organizations set minimum values for the factor of safety. For example, the standard for lifting hoists and elevators require a minimum FOS of 4, because their failure would involve the clear risk of injuring or killing people. As a guide, consider the FOS as a product of factors: $\text{FOS} = \prod_{k=1}^{n} F_k = F_1 F_2 F_3 \ldots F_n$. Table 3.1 gives a set of typical factors to be considered.

3.10. Element Type Selection

Even with today's advances in computing power you seem never to have enough computational resources to solve all the problems that present themselves. Frequently solid elements are not the best choice for computational efficiency. The analysts should learn when other element types can be valid or when they can be utilized to validate a study carried out with a different element type. SW Simulation offers a small element library that includes bars, trusses, beams, frames, thin plates and shells, thick plates and shells, and solid elements. There are also special connector elements called rigid links or multipoint constraints. They are mathematical simplifications used to connect different points, curves, surfaces, or bodies together.

Table 3.1. Factors to consider when evaluating a design (each ≥ 1).

k	Type	Comments
1	Consequences	Will loss be okay, critical or fatal
2	Environment	Room-ambient or harsh chemicals present
3	Failure theory	Is a part clearly brittle, ductile, or unknown
4	Fatigue	Does the design experience more than ten cycles of use
5	Geometry of part	Not uncertain, if from a CAD system
6	Geometry of mesh	Defeaturing can introduce errors. Element sizes and location are important. Looking like the part is not enough
7	Loading	Are loads precise or do they come from wave action, etc.
8	Material data	Is the material well known, or validated by tests
9	Reliability	Must the reliability of the design be high
10	Restraints	Designs are greatly influenced by assumed supports
11	Stresses	Was stress concentration considered, or shock loads

The shells and solid elements are considered to be continuum elements. The plate element is a special case of a flat shell element with no initial curvature. Solid element formulations include the stresses in all directions. Shells are a mathematical simplification of solids of special shape. Thin shells (like thin beams) do not consider the stress in the direction perpendicular to the shell surface. Thick shells (like deep beams) can consider the stresses through the thickness on the shell, in the direction normal to the middle surface, and account for transverse shear deformations.

Let h denote the typical thickness of a component while its typical length is denoted by L. The thickness to length ratio, h/L, gives some guidance as to when a particular element type is valid for an analysis. When h/L is large shear deformation is at its maximum importance and you should be using solid elements. Conversely, when h/L is very small transverse shear deformation is not important and thin shell elements are probably the most cost effective element choice. In the intermediate range of h/L the thick shell elements will be most cost

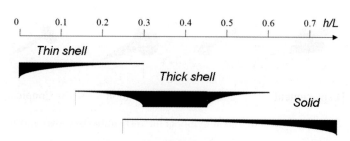

Fig. 3.9. Overlapping valid ranges of element types.

effective. The thick shells are extensions of thin shell elements that contain additional strain energy terms.

The overlapping h/L ranges for the three continuum element types are suggested in Figure 3.9. The thickness of the lines suggests those regions where a particular element type is generally considered to be the preferred element of choice. The overlapping ranges suggest where one type of element calculation can be used to validate a calculated result obtained with a different element type. Validation calculations include the different approaches to boundary conditions and loads required by different element formulations. They also can indirectly check that a user actually understands how to utilize a finite element code.

3.11. Simulation Fixture and Load Symbols

The symbols used in SW Simulation to represent a single translational and rotational DOF at a node are shown green in Figure 3.10. The symbols for the corresponding forces and moment loadings are shown pink in that figure. Since finite element solutions are based

Node of solid or truss element: Node of frame or shell element:
All three displacements are zero.   Zero displacements and rotations.

Fig. 3.10. Fixed restraint symbols for solids (left) and shell nodes.

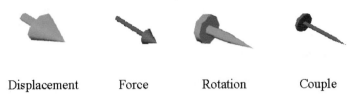

Displacement　　　Force　　　Rotation　　　Couple

Fig. 3.11.　Single component symbols for restraints (fixtures) and loads.

on work–energy relations, the above word "corresponding" means that their dot product represents the mechanical work done at the point. When a model can involve either translations or rotations as DOF they are often referred to as generalized displacements. The SW Simulation nodal symbols for the unknown generalized displacement DOF's for the solid nodes (top) and shell nodes are seen in Figure 3.11. You almost always must supply enough restraints to prevent any model from undergoing a rigid body translation or rigid body rotation.

For simplicity many finite element examples incorrectly apply complete restraints at a face, edge or node. That is, they enforce an **Immovable** condition for solids or a **Fixed** condition for shells. Actually determining the type of restraint, as well as where the part is restrained is often the most difficult part of an analysis. You frequently encounter the conditions of symmetry or anti-symmetry restraints. You should under understand symmetry plane restraints for solids and shells.

A plane of symmetry is flat and has mirror image geometry, material properties, loading, and restraints. Symmetry restraints are very common for solids and for shells. Figure 3.12 shows that for both solids and shells, the displacement perpendicular to the symmetry plane is zero. Shells have the additional condition that the in-plane component of its rotation vector is zero. Of course, the flat symmetry plane conditions can be stated in a different way. For a solid element translational displacements parallel to the symmetry plane are allowed. For a shell element rotation is allowed about an axis perpendicular to the symmetry plane and its translational displacements parallel to the symmetry plane are also allowed.

Node of a solid or truss element: Displacement normal to the symmetry plane is zero.

Node of a frame or shell element: Displacement normal to the symmetry plane and parallel rotations are zero.

Fig. 3.12. Symmetry: zero normal displacement, and in-plane rotation.

Table 3.2. Fixtures for solid stress analysis.

Fixture type	Description
Circular symmetry	Periodically repeated segments have the same unknown displacements.
Fixed	All three translations and rotations are zero on face, edge, or vertex.
Fixed hinge	On a cylindrical face, only the circumferential displacement is allowed.
Immovable	All three translations are zero on face, edge, or vertex.
On cylindrical faces	The cylindrical coordinate displacements normal to and/or on the cylindrical surface are given.
On flat faces	Displacements normal to and/or tangent to a flat face.
On spherical faces	The spherical coordinate displacements normal to and/or on the spherical surface are given.
Roller/sliding or symmetry	Two displacements tangent to a flat face are allowed.
Use reference geometry	A face, edge, or vertex can translate a specified amount relative to a reference plane and axis.

3.12. Structural Restraint Options in SolidWorks

Solids and shells must be restrained and loaded in different ways since shells also have rotational degrees of freedom and solids do not. Table 3.2 lists the current restraint or fixture options for solids within SW Simulation. Note that they only involve components of the displacement vectors. The loading options for solid parts are given in Table 3.2. No pure couples (moments) can be specified at a node.

3.13. Structural Connectors

In an FEA it is common to add linear constraint equations to the matrix system. They are often called multipoint constraints (MPC) or connectors. They can be converted to symmetric matrix identities that resemble finite element matrices, and therefore can be assembled into the system of equations to be solved. Alternatively, they can be enforced as Lagrange multipliers that increase the size of the matrix equations. For example, assume to points displace such that there is a constant gap size between them. You can write $u_1 - u_2 = gap$. That is just a special case of a general linear constraint such as $u_m + c_n u_n = c_m$. Since the equation involves multiple DOF (usually at different points) it developed the name of a multipoint constraint. When you understand the mechanics of a particular type of physical connection you can approximate it as a set of MPC equations and name it as if it were the physical part. That lets you avoid modeling the physical parts of such a connection, at least in the preliminary stages of a study. Table 3.3 gives the available list of time independent

Table 3.3.   Time independent structural connectors.

Connector	Description
Bearing	Simulates interaction between a cylindrical shaft surface and a support bearing.
Bolt	Simulates a connection between multiple components, or a component and ground.
Elastic support	Defines an elastic (Winkler) foundation between a face and the ground.
Link	Ties the translation of two vertices rigidly together (but not any rotations).
Pin	Connects cylindrical faces or circular edges together (rigidly or with stiffness's).
Rigid	Rigidly connects two faces together so the distance between them is held constant.
Spot weld	Rigidly connects two small shell faces together (same displacement and rotation).
Spring	Linear spring between center of two faces (general, tension only, compression only).

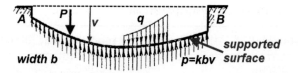

Fig. 3.13. Winkler foundations deflect only under the contact surface.

connectors. For time dependent problems a spring-damper connector is available between two vertices.

The elastic support connector is also referred to as an elastic foundation, a virtual wall, or a Winkler foundation. It is a common type of structural support approximation. It assumes that the omitted support applies an opposing pressure, from linear springs, that is directly proportional to the displacement component normal (and/or tangential) to the assumed contact surface. In the mathematical sense, it is called a mixed boundary condition. The other end of the foundation spring system is taken to be fixed to a rigid base. As a result, an elastic support prevents rigid body translation in the direction of the support. It also prevents rigid body rotation about an axis lying is a support plane. As sketched in Figure 3.13, a Winkler foundation only deforms the support material directly in contact with the part, but not the adjacent support material. The opposing pressure is usually compressive, but it can become tensile.

The foundation constant of proportionality, k_0, is called the support stiffness or foundation stiffness. It is defined as the force per unit area (pressure) required to produce a unit deflection of the elastic support. Thus, its units can be thought of as $(N/m^2)/m$ or N/m^3 or $(N/m)/m^2$, or similar English units. The first choice follows directly from its definition. The last choice follows from a related approximation where the foundation is assumed to be sitting on a large number of discrete springs, of stiffness units (N/m), that each supports a small sub-area (m^2) of the foundation. Often the support stiffness is determined by experimentation, especially for soils. Some typical values are listed in Table 3.4. Large values of k_0 make the support act as if it is completely fixed in the stiffness direction. If the supporting foundation is a single layer of known elastic material

Table 3.4.   Typical elastic foundation modulus ranges.

Material	Foundation modulus, k_0, $[\text{N/m}^3]$
Concrete	$270e6$
Hard rubber	$200\text{–}350e6$
Railroad ballast	$200e6$
Soil, compacted	$108\text{–}120e6$
Soil, loose	$29\text{–}40e6$
Wood	$100e6$

then the normal stiffness can be taken as $k_0 = E/t$, where E is its elastic modulus and t is the thickness. Replace E with G, the shear modulus of the material, to specify a tangential foundation support. If the foundation is made up of multiple layers of elastic materials then the normal foundation stiffness is $k_0 = [\sum_j E_j/t_j]^{-1}$.

3.14. Available Structural Loading Options

Most finite element systems have a wide range of mechanical loads (or sources) that can be applied to points, curves, surfaces, and volumes. The mechanical loading terminology used in SW Simulation is in Table 3.5. Most of those loading options are utilized in later example applications.

Table 3.5.   Structural loads that apply to the active structural study.

Load type	Description
Bearing load	Non-uniform bearing load on a cylindrical face
Centrifugal force	Radial centrifugal body forces for the angular velocity and/or tangential body forces from the angular acceleration about an axis
Force	Resultant force, or moment, at a vertex, curve, or surface
Gravity	Gravity, or linear acceleration, body force loading
Pressure	A normal and/or tangential pressure acting on a selected surface
Remote load/mass	Allows loads or masses remote from the part to be applied by treating the omitted material as rigid
Temperature	Temperature at selected curves, surfaces, or bodies (see thermal studies)

Table 3.6. Isotropic mechanical properties.

Symbol	Label	Item
E	EX	Elastic modulus (Young's modulus)
μ	NUXY	Poisson's ratio
G	GXY	Shear modulus
ρ	DENS	Mass density
σ_t	SIGXT	Tensile strength (ultimate stress)
σ_c	SIGXC	Compression stress limit
σ_y	SIGYLD	Yield stress (yield strength)
α	ALPX	Coefficient of thermal expansion

3.15. Available Material Inputs for Stress Studies

Most applications involve the use of isotropic (direction independent) materials. The available mechanical properties for them in SW Simulation are listed in Table 3.6. It is becoming more common to have designs utilizing anisotropic (direction dependent) materials. Their input options are listed in Table 3.7.

The most common special case of anisotropic materials is the orthotropic material. Any anisotropic material has its properties input relative to the principal directions of the material. That means you must construct the principal material directions reference plane or coordinate axes before entering orthotropic data. Mechanical orthotropic properties are subject to some theoretical relationships that physically possible materials must satisfy (such as positive strain energy). Thus, experimental material properties data may require adjustment before being accepted by SW Simulation.

3.16. Stress Study Outputs

A successful run of a study will create a large amount of additional output results that can be displayed and/or listed in the post-processing phase. Displacements are the primary unknown in a SW Simulation stress study. The available displacement vector components are cited in Tables 3.8 and 3.9, along with the reactions they create if the displacement is used as a restraint.

Table 3.7. Anisotropic mechanical properties in principal material direction.

Symbol	Label	Item
E_x	EX	Elastic modulus in material X direction
E_y	EY	Elastic modulus in material Y direction
E_z	EZ	Elastic modulus in material Z direction
μ_{xy}	NUXY	Poisson's ratio in material XY directions
μ_{yz}	NUYZ	Poisson's ratio in material YZ directions
μ_{xz}	NUXZ	Poisson's ratio in material XZ directions
G_{xy}	GXY	Shear modulus in material XY directions
G_{yz}	GYZ	Shear modulus in material YZ directions
G_{xz}	GXZ	Shear modulus in material XZ directions
ρ	DENS	Mass density
σ_t	SIGXT	Tensile strength (ultimate stress)
σ_c	SIGXC	Compression stress limit
σ_y	SIGYLD	Yield stress (yield strength)
α_x	ALPX	Thermal expansion coefficient in material X
α_y	ALPY	Thermal expansion coefficient in material Y
α_z	ALPZ	Thermal expansion coefficient in material Z

Table 3.8. Output results for solids, shells, and trusses.

Symbol	Label	Item	Symbol	Label	Item
U_x	UX	Displacement (X direction)	R_x	RFX	Reaction force (X direction)
U_y	UY	Displacement (Y direction)	R_y	RFY	Reaction force (Y direction)
U_z	UZ	Displacement (Z direction)	R_z	RFZ	Reaction force (Z direction)
U_r	URES	Resultant displacement magnitude	R_r	RFRES	Resultant reaction force magnitude

The displacements can be plotted as vector displays, or contour values. They can also be transformed to cylindrical or spherical components. The strains and stresses are computed from the displacements. The stress components available at an element centroid or averaged at a node are given in Table 3.10. The six

Table 3.9. Additional primary results for beams, plates, and shells.

Symbol	Label	Item	Symbol	Label	Item
θ_x	RX	Rotation (X direction)	M_x	RMX	Reaction moment (X direction)
θ_y	RY	Rotation (Y direction)	M_y	RMY	Reaction moment (Y direction)
θ_z	RZ	Rotation (Z direction)	M_z	RMZ	Reaction moment (Z direction)
			M_r	MRESR	Resultant reaction moment magnitude

Table 3.10. Nodal and element stress results.

Sym	Label	Item	Sym	Label	Item
σ_x	SX	Normal stress parallel to x-axis	σ_1	P1	1st principal normal stress
σ_y	SY	Normal stress parallel to y-axis	σ_2	P2	2nd principal normal stress
σ_z	SZ	Normal stress parallel to z-axis	σ_3	P3	3rd principal normal stress
τ_{xy}	TXY	Shear in Y direction on plane normal to x-axis	τ_I	INT	Stress intensity twice the maximum shear stress
τ_{xz}	TXZ	Shear in Z direction on plane normal to x-axis			
τ_{yz}	TYZ	Shear in Z direction on plane normal to z-axis	σ_{vm}	VON	von Mises stress (distortional energy failure criterion)

components listed on the left in that table give the general stress at a point (i.e., a node or an element centroid). Those six values are illustrated on the left of Figure 3.14. The corresponding strains available for output at the element centroids are listed in Table 3.11. Stresses can be used to compute the scalar von Mises failure criterion.

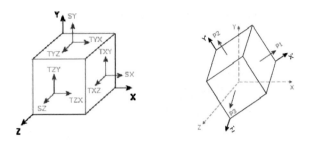

Fig. 3.14.　Stress tensor (left) and its principal normal values.

Table 3.11.　Element centroidal strain component results.

Sym	Label	Item	Sym	Label	Item
ε_x	EPSX	Normal strain parallel to x-axis	ε_1	E1	Normal principal strain (1st principal direction)
ε_y	EPSY	Normal strain parallel to y-axis	ε_2	E2	Normal principal strain (2nd principal direction)
ε_z	EPSZ	Normal strain parallel to z-axis	ε_3	E3	Normal principal strain (3rd principal direction)
γ_{xy}	GMXY	Shear strain in Y direction on plane normal to x-axis	ε_r	ESTRN	Equivalent strain
γ_{xz}	GMXZ	Shear strain in Z direction on plane normal to x-axis	SED	SEDENS	Strain energy density (per unit volume)
γ_{yz}	GMYZ	Shear strain in Z direction on plane normal to y-axis	SE	ENERGY	Total strain energy

They can also be used to solve an eigenvalue problem for the principal normal stresses and their directions, which are shown on the right of Figure 3.14. The maximum shear stress occurs on a plane whose normal is 45 degrees from the direction of P1. The principal normal stresses can also be used to compute the von Mises failure criterion. It is a positive scalar having the units of stress, but is a measure of the distortional strain energy.

The von Mises effective stress is compared to the material yield stress for ductile materials. Failure is predicted to occur (based on the distortional energy stored in the material) when the von Mises value reaches the yield stress. The maximum shear stress is predicted to cause failure when it reaches half the yield stress. SW Simulation uses the shear stress intensity which is also compared to the yield stress to determine failure (because it is twice the maximum shear stress). The first four values on the right side of Table 3.10 are often represented graphically in mechanics as a 3D Mohr's circle.

If desired, you can plot all three principal components at once. The three principal normal stresses at a node or element center can be represented by an ellipsoid. The three radii of the ellipsoid represent the magnitudes of the three principal normal stress components, P1, P2, and P3. The sign of the stresses (tension or compression) are represented by arrows. The color code of the surface is based on the von Mises value at the point, a scalar quantity. If one of the principal stresses is zero, the ellipsoid becomes a planar ellipse. If the three principal stresses have the same magnitude, the ellipsoid becomes a sphere. In the case of simple uniaxial tensile stress, the ellipsoid becomes a line.

The available nodal output results in Table 3.10 are obtained by averaging the element values that surround the node. You can also view them as constant values at the element centroids. That can give you insight to the smoothness of the approximation. For brittle materials you can also be interested in the element strain results, listed in Table 3.11.

3.17. Stress Concentration and Defeaturing

An important concept in stress analysis is stress concentration. Small local geometric feature changes in a part, such as a fillet, hole or notch, can significantly change the stress magnitudes at the boundary of the small geometric feature. A useful visualization is to imagine an

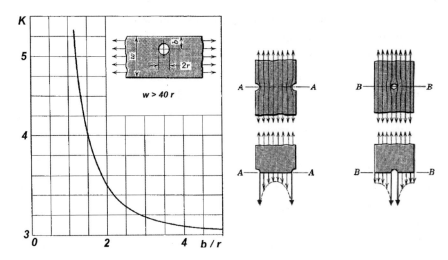

Fig. 3.15.   An edge hole significantly increases the maximum stress.

ideal fluid flowing through the interior of the part: initially without the feature, and again with the feature. The disturbance of the streamlines is analogous to the local disruption in the local (nominal) stress field. Such stress flow lines are shown in Figure 3.15 where the local constant (or linear) nominal stress field changes its spatial distribution and develops a local maximum value. Generally, the maximum local stress increases logarithmically as the size of the feature decreases. That is, the concentration factor versus a geometric size ratio is a straight line on a log–log plot. For most common features, the scaling "Stress Concentration Factor", K, has been determined from elasticity theory, FEA studies, and experimental studies. Having access to tabulated stress concentration graphs can aid in validating FEA study results.

It is common for the magnitude of the maximum local stress to increase by a factor of three, or more. For example, if a small hole is in the center of a large plate with a one-dimensional stress, then the stress concentration factor is exactly 3.0. However, if the same hole is moved near on edge of the plate the concentration factor rapidly increases above five (see Figure 3.15). How the local stress distribution changes is less important than the new local maximum magnitude defined as $\sigma_{max} = K\sigma_{nom}$ where K is the stress concentration factor, and σ_{nom} is the nominal local stress. The nominal stress is the local stress calculated on the assumption that the feature does not increase the stress. The nominal stress was typically taken as a constant or linear variation as found from an analytic approximation, but today might be obtained by an FEA.

At times you will be tempted to defeature a small region in a part because it creates so many small elements that it greatly slows the execution time and/or causes the study to run out of memory. Before accepting a defeature operation you should consider the estimated local nominal stress field and the stress concentration factor for a similar geometry.

3.18. Classic 1D Analytic Stress Solutions

In a typical study, you execute a complex FEA study and then seek simplified solutions (or a simplified FEA model) in an attempt to validate your study. When you start working with new software it is wise to reverse the usual process and run a problem for which the results are known. That lets you be sure you understand the proper utilization of the software. There are a few well know 1D stress analysis problems that have simple solutions that give you insight into structural solutions and are easily verified with a SW Simulation analysis. An axial bar subject to a constant end load has a deflection that is linear and the stress is constant. Thus, any FEA gives the exact result everywhere for both quantities. Consider a bar with two types of distributed loads that test different options in an FEA system: gravity and acceleration due to an angular velocity.

Let the bar be supported at one end. In the first case, it simply hangs from that end loaded with its own weight. In the second case, the bar rotates about that end at a constant 1,000 rpm. The bar has a cross-section of 0.1 m squared and a length of a half meter $(A = 0.01\,\text{m}^2, L = 0.5\,\text{m})$. The material is a high lead content bronze $(E = 1.1\text{e}11\,\text{Pa}, \nu = 0.33, \rho = 8,700\,\text{kg/m}^3)$. The one-dimensional model does not consider the effect of Poisson's ratio. For the gravity load, the 1D approximate axial displacement (UX) and stress (SX) are

$$u(x) = \frac{\gamma A L^2}{E} \left[\frac{1}{2} \left(\frac{x}{L} \right)^2 - \frac{x}{L} \right], \quad \sigma(x) = \gamma L \left(\frac{x}{L} - 1 \right), \quad \gamma = \rho g.$$

A single quadratic element would yield the exact solution everywhere (for no Poisson effect). This part is constructed and a static simulation study is opened.

(1) The end plane is supported on rollers and the centerline of that plane is fixed against translation to eliminate RBM via **Fixtures → Roller/Slider** and then **Fixed Geometry**.

(2) The gravity load is activated by selecting a horizontal plane (Figure 3.16) to be normal to the gravity vector with **External Loads → Gravity**. The end (right) plane is selected and the gravitational acceleration is set to $9.81 \, \text{m/s}^2$.

(3) A crude default mesh is generated and the problem is executed to create the results via: **Mesh → Create Mesh, OK** and **Study Name → Run**.

(4) The displacement results are checked first: **Results → Define Displacement Plot → UX**. The contour and graph of the displacement matches the 1D approximation very closely (except at support centerline). The graph is produced by a right click on the plot name, **List Selected → Edge Line → Update → Graph**. The mesh and displacement graphics are given in Figure 3.17.

(5) The axial stress results (Figure 3.18) are checked first: **Results → Define Stress Plot → SX**. The contour and graph of

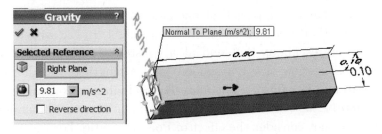

Fig. 3.16.　Apply a gravitation acceleration to the part volume.

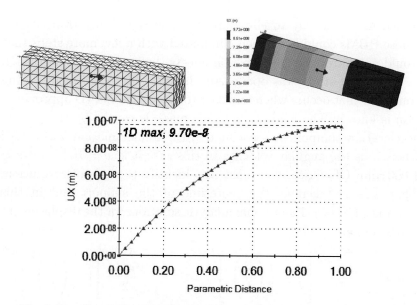

Fig. 3.17. Gravity loading displacements match the 1D approximation.

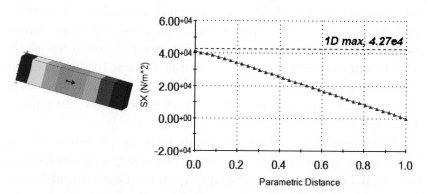

Fig. 3.18. Gravity loading stress matches the 1D approximation.

the stress matches the 1D approximation very closely (except at support centerline).

The 1D approximation matches very well, except along the fixed centerline of the end plane. There false Poisson ratio effects are introduced into the solid part since points on that line are prevented from contracting as they would in a physical part. That restraint was picked as a fast way to eliminate in-plane rigid body translations. Those RBMs could have been eliminated with a few more steps that would not introduce false Poisson effects (how would you do that?).

A second study was opened to illustrate the centrifugal loading condition that occurs when any part rotates. Again, a 1D approximation is known. The loading, per unit mass, is caused by the normal acceleration component associated with circular motion: $a_n = r\omega^2$, where ω is the angular velocity. In this case it is $\omega = 104.7 \,\mathrm{rad/s} = 1{,}000 \,\mathrm{rpm}$. Of course, you must also identify the axis of revolution. That axis is taken as the centerline of the support end in this example. The one-dimensional analytic solutions for the displacement and stress are:

$$u(x) = \frac{\rho\omega^2 L^3}{2E}\left[\frac{x}{L} - \frac{1}{3}\left(\frac{x}{L}\right)^3\right], \quad \sigma(x) = \frac{\rho\omega^2 L^2}{2}\left[1 - \left(\frac{x}{L}\right)^2\right].$$

A cubic response like this cannot be modeled by a single quadratic bar element, so the results are now mesh dependent. A mesh of quadratic elements will approximate the parabolic stress versus position curve as a series of straight line segments. In other words, the element stresses are piecewise linear from element to the next, and discontinuous at the element interfaces. Surprisingly, the quadratic element nodal displacements will be exact at the nodes, but approximate interior to the element.

The restraint procedure and result recovery procedure are the same as the prior study. The centrifugal loading is applied via **External Loads → Centrifugal** to open the **Centrifugal Panel**. There select the axis of revolution and specify the angular velocity value, as seen in Figure 3.19. The angular acceleration option was not used because a constant rotational speed was given.

The axial displacement (Figure 3.20) and axial stress results (Figure 3.21) match the cubic approximation very closely. The lateral displacements in the solid (not shown) are due to Poisson ratio effects only, and the lateral stresses (not shown) are almost zero except near the axis of revolution. The restrained lateral displacements there create false stresses due to the modeling approach discussed above.

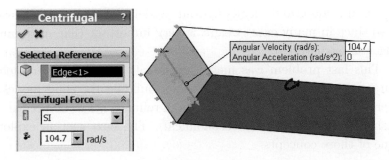

Fig. 3.19. Applying the centrifugal loading about a fixed axis.

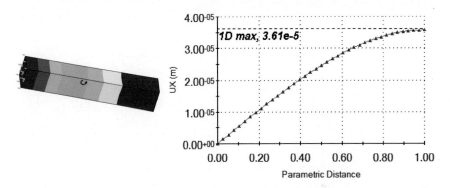

Fig. 3.20. Displacement matches cubic 1D approximation very closely.

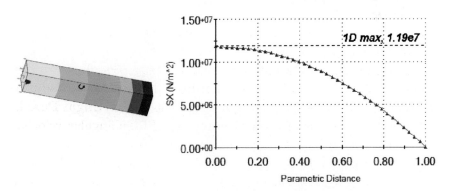

Fig. 3.21. Axial stress matches 1D quadratic approximation very closely.

Comparing the two load cases, you see that even a relatively low rotation speed can cause stress levels that exceed those due to gravity by a factor of 1,000 or more. That is one reason weight is sometimes initially neglected in machine design. However, it is so easy to include gravity loads in your model that you should always do so since in many cases weight is very important (like medium to long span structures).

This last problem was mesh dependent, as are most problems. You need to learn to plan ahead when building the solid model to insert split lines where you expect mesh refinements to be needed (or mesh de-refinements to be economical). The next section explores some of those concepts.

4

Mesh Control in SolidWorks Simulation

4.1. Introduction

You must plan ahead when building a solid model so that it can be used for realistic finite element load and/or restraint analysis cases. You often do that in the solid modeling phase by using lines or arcs to partition lines, curves, or surfaces. This is called using a "split line" in SolidWorks and SW Simulation. There is also a "split part" feature that is similar except that it cuts a part into multiple sub-parts. You sometimes need to do that for symmetry or anti-symmetry finite element analysis so that we can analyze the part more efficiently. Here, the concept of splitting surfaces for mesh control will be illustrated via stress analysis.

4.2. Example Initial Analysis

The split line concepts for mesh control will be illustrated via the first tutorial "Static Analysis of a Part", using the SW Simulation example file Tutor1.sldprt which is shown in Figure 4.1. (Remember to save it with a new name by putting your initials at the front.) Consider that tutorial to be a preliminary analysis. You should recall that there was bending of the base plate near the loaded vertical post. However, the original solid mesh had only one element through the thickness in that region and would therefore underestimate the bending stresses there. You need to control the mesh there to

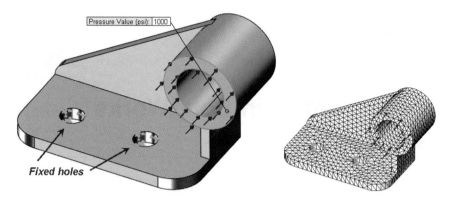

Fig. 4.1.   Original part, restraints, load, and mesh.

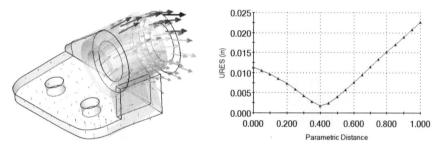

Fig. 4.2.   Original part vector deflections and bottom right edge graph.

form 4 to 5 layers to accurately capture the change in bending stress through the thickness.

The original deformed shape, in Figure 4.2, is shown relative to the undeformed part (in gray). You see bending at the end of the base and deflection of the part bottom back edge in the direction of the (unseen) supporting object below it. From the original effective stress plot in Figures 4.3 and 4.4 you can see that large regions, within the red contours, have exceeded the material yield stress. Actually, the maximum value is over 133,000 psi, or about 1.5 times the yield stress, on the top and bottom of the base. Clearly, this part must have its material or dimensions changed and/or new support options must be utilized.

Begin a revision of the first study of this part by reconsidering the restraints utilized. In many problems the restraints are unclear

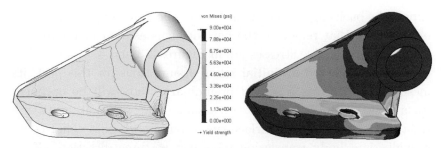

Fig. 4.3.   Original top surface effectives stresses.

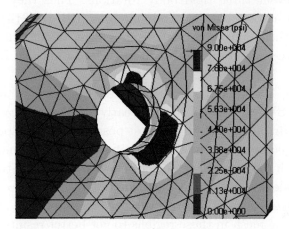

Fig. 4.4.   Original bottom surface effective stresses.

or questionable and you need to consider other restraint cases. You previously restrained all the three translations on the two small cylindrical surfaces. That makes those surfaces perfectly rigid. That would be almost impossible to build. It is likely that the holes were intended to be bolt holes. Then the applied backward $(-z)$ pressure load would probably require the development of tension reaction forces along the back $(-z)$ half of the bolt cylinder. That would not happen because an air gap would open up. Also, a bolt usually applies a restraint to the surface under the bolt head (in addition to a bolt bearing load on its cylindrical part), and that is not in the original choice of restraints. Furthermore, the bracket seems to be attached to some rigid object under it and therefore you would expect the back edge of the bracket to be somehow supported by that object when that region of the part deforms as seen in Figure 4.2. The original effective stresses are in Figures 4.3 and 4.4.

Having reviewed an initial set of assumptions for this part, its mesh and its results a second preliminary study will be outlined. As a new possible restraint set, assume that the bolt heads are tight and act on a small surface ring around each hold. That could provide rigid body translation restraints in three directions. Each bolt would prevent three translations and the pair of them combines to also prevent three rigid body rotations. Thus they combine to prevent all six possible rigid body motions (RBM). If the bolts were not tight then only a normal displacement (along the bolt axis) would be restrained on the base top surface (and two RBM would remain).

Bearing loads on the bolt shaft can be found by an iterative process in SW Simulation, but for early studies you can assume a small cylindrical contact at the most positive z (front) location. The previously computed bending of the part is also assumed to cause contact with the supporting object below, and thus a y-translational restraint, along the bottom back edge of the part. To accomplish those types of restraint controls you need to "split" the surfaces and transition the mesh in those regions to get better results. Thus, you need to form two smaller surface rings for the bolt heads, and split the cylinders into smaller bearing areas. That is done by adding split lines to selected regions of the part.

4.3. Splitting a Surface or Curve

To avoid changing the master tutorial file, open the part and then "save as" a new file name on the desktop. To introduce the required split lines:

(1) Select the top surface of the base by moving the cursor over it until its boundary turns green and **Insert Sketch**. Pick the **Sketch** icon and pick the **Circle** option to form a bolt head washer area.

(2) Place the cursor on the hole edge to "wake up" the center point. Draw a larger circle on the surface. Set its diameter to 30 mm and click **OK**. That does not change the surface; it just adds a circle to it.

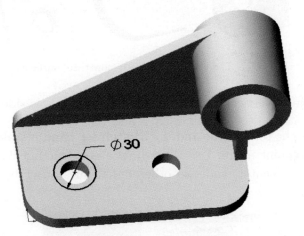

(3) To split the surface, go to the top and select **Insert** → **Curve** → **Split Line**. The **Type** will be a projection.

(4) Next, pick the surface(s) to be cut by this curve. Here it will be the top of the base plate again so select it and click **OK**. Repeat the above two processes for the second hole.

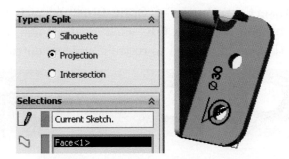

Now you will see that moving the cursor around shows a new circle and a new ring of surface area that could be used to enforce restraints or loads. The new surface areas are shown in Figure 4.5. Later you will use these newly created ring areas as a bolt head (washer) restraint region.

Fig. 4.5.   The original surface is now three surfaces.

To form a vertical bearing region on the bolt hole sidewall:

(1) Select the top surface and use **Sketch** → **Line**. "Wake up" the center point again. Use it to draw a construction line forward from the center, and two other radial lines offset by about 25 degrees.

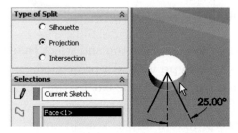

(2) Select **Insert** → **Curve** → **Split Line** and pick the cylinder of the first bolt hole, click **OK**. That creates a new load bearing surface that could be used to restrain and/or control the mesh.

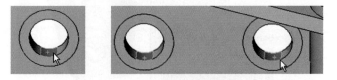

(3) Repeat those operations for the second bolt hole.

The above surface splits were constructed to give more flexibility in applying various displacement restraints. You will need another surface split to help exercise engineering control over the revised mesh. You want the "L" shaped side area to have the leg split off so you can control a bending mesh there:

(1) Right click on the $(+x)$ right side face, **Insert Sketch** $\rightarrow$ **Sketch** $\rightarrow$ **Line**.

(2) "Wake up" the left vertical line to put in a short line that crosses the base.

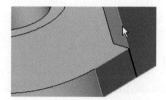

(3) Use **Insert** $\rightarrow$ **Curve** $\rightarrow$ **Split Line** and select the "L" face, click **OK**. Now it has become two rectangles.

All of these new surface areas and lines can have different local mesh sizes prescribed to control our mesh generation. That is a standard feature of SW Simulation, but you must supply the engineering judgment as to where any part needs the extra line or surface divisions for applying loads (or heat sources) and restraints in an analysis. True point loads or point moments are unusual in practice. They should be replaced by reasonable loading areas (that require split curves) and pressure distributions.

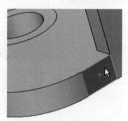

4.4. Beginning SW Simulation Study

Activate the SW Simulation Manager by clicking on its icon:

(1) Then start a static study. In this case the element type defaults to a quadratic solid. **Part name** $\rightarrow$ **Apply/Edit Material**. Pick alloy steel as the material.

(2) Note that you have a choice of display units for the properties. The number of significant figures in the test data should be independent of the units selected. Any difference is due to using a higher number of digits in the units conversion. Do not be mislead by a long string of digits in materal values. You can usually trust the first three, or maybe four. Note for later use that the yield stress of this material is about 620 MPa or 90 ksi.

4.5. Mesh Control

Wherever translational displacement restraints are applied, reaction forces are developed and localized stress concentrations are likely. Therefore, you want to assure that small elements are created in such regions. This process is referred to as mesh control. It is required in almost every study. Invoke it with:

(1) Right click on **Mesh → Apply Control** to bring up the **Mesh Control panel**. The default element size of about 3 mm needs to be changed in the new regions created above. Select the region around the two bolt washers by picking those two surfaces, and set the desired size to 1 mm.

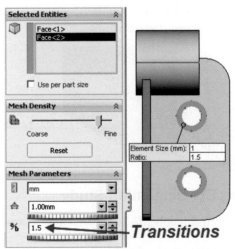

(2) Likewise, for the two bearing surfaces within the smaller vertical cylindrical holes set the size to about 1.0 mm.

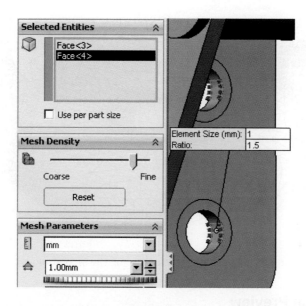

(3) In the final corner region you need several elements through the thickness of curved front corner to accurately model the local bending seen in the first study. Select the small split rectangle and the adjacent cylindrical corner and use an element size of about 1.5 mm.

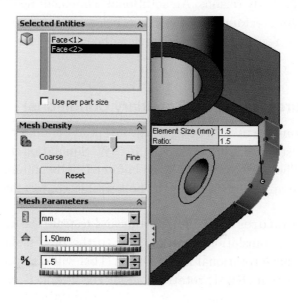

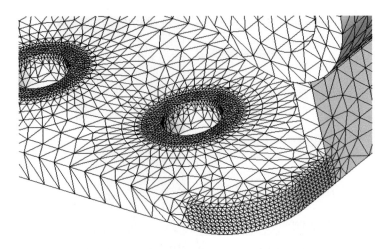

Fig. 4.6. Controlled mesh sizes for the second model.

4.6. Mesh Preview

Use **Mesh → Create** to generate the mesh. Then examine the mesh and increase or decrease the local sizes specified above so that it looks acceptable, as shown in Figure 4.6. Note that the mesh makes a smooth transition from the smaller element sizes to the larger default size in the far body regions. An additional refinement near the corner of the base and rectangular shaped leg (below the loaded tube) would also be wise. You should always preview the mesh before running the solution.

4.7. Fixtures (Essential Boundary Conditions)

Various terms are used to describe the essential boundary conditions: fixtures, restraints, supports, etc. In SW Simulation the term is Fixtures. The part fixtures will be enforced by beginning with the new surface areas representing the bolt washer contact regions:

(1) Select **Fixtures → Roller/Slider** and restrain the washer areas against y-translation perpendicular to the base (solid elements do not have rotational nodal DOF). This will also prevent rigid body motion (RBM) rotations about the x- and z-axes.

(2) In the **Standard panel** pick the two concentric ring faces as the **Selected Entities** and chose **Roller/Slider** as the **Type**.

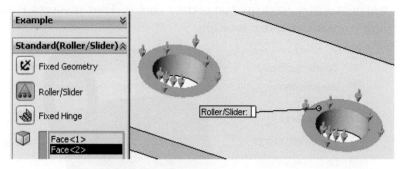

(3) Next select the two cylindrical bolt bearing faces and **restrain radial translation** on the **Advanced Fixtures → On Cylindrical Faces Type**. That prevents x- *and* z-translational RBM and the y rotational RBM. At least all RBMs are now safely accounted for.

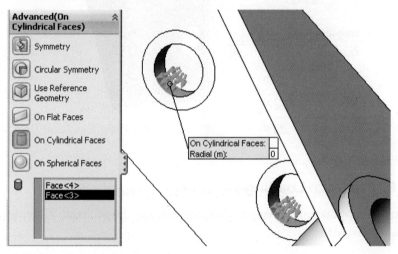

(4) Finally, this study will assume the bottom back edge helps support the part due to contact with the strong background material, to which the bolts are also *assumed* to be rigidly attached. Use **Fixtures → Advanced → Use Reference Geometry**. Select that back edge line and choose the bottom plane as your **reference plane Type**. Restrain the edge line

displacement, relative to the bottom, by picking the vertical direction (perpendicular to the bottom plane) and preview the restraint arrow to verify the correct choice. That prevents RBM in y translation and z rotation.

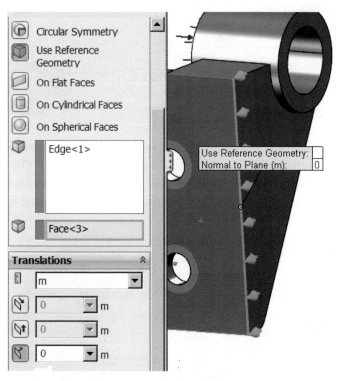

4.8. Pressure Loading

This part has one pressure loading on the large front tube face. Impose it with:

(1) **External Loads** → **Pressure**. Give the **Pressure Type** as normal to the selected face.
(2) Set the units and value (1,000 psi) and preview the arrows to verify the direction (sign of the pressure), **OK**. At this point, you may wish to take advantage of the standard Windows feature and rename this load condition from the default *Pressure_1* to

Front_ring_pressure, by doing a slow double click on the default name.

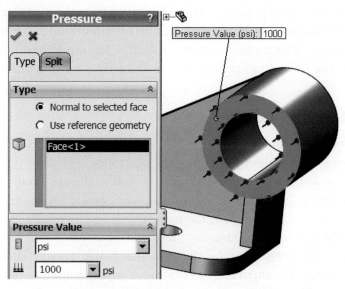

Clearly, the resultant applied force will be that pressure times the selected surface area (about 2,840 lb). Since SolidWorks is a parametric modeler remember that if you change either diameter of the tube the area and the resultant force will also change. If you mean to specific the total force then use **External Loads → Force** and give the total force on that area. It would not change with a parametric area change.

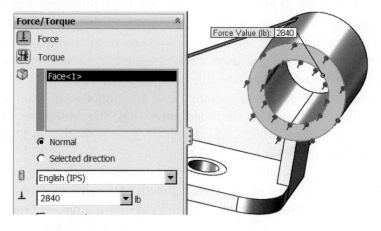

Remember that when the study results are obtained later, to check the total force caused by the above pressure you can check the reaction force since it must be equal and opposite. (You can do that in post-processing by right clicking on the **Displacement report** → **Reactions**). The z-reaction force will give the equal and opposite force to the total applied here (this is the only z-force applied).

4.9. Run the Study

Right click on the study name and select **Run**. Proceed to post-processing the results by checking the reactions, reviewing the "load path", plot the displacement vectors and contours, and display selected stress (or strain) components, etc. Upon completion, try to validate the results and plan the next revision or "what-if" question about the part and its intended function.

After any structural or thermal study you should check the reactions from the surroundings that were approximated as restraints or fixtures. They will be equal and opposite to the resultants of the applied loads or heat sources. Right click on **Results** → **List Result Force** → **Reaction force**. Under **Selection** pick the restraint regions in the model, and set the units. **Update** gives the plot shown (Figure 4.7).

The ring of pressure applied in this problem was totally in the z-direction. Therefore, there should be no reaction forces in the x- and y-directions. While not zero they are 10^{-4} times smaller. The iterative solver was used to get the current results. It is much faster than the optional direct solver, but a direct solve will be slightly more accurate. When obtained from a direct solve the reaction forces are more correct (below), but the difference is not important.

Some engineers try to visualize how the major loads are transferred to the supports via the "load path". SW Simulation has a tool to assist with that (Figure 4.8): Right click on **Results** → **Define Design Insight Plot**.

Start with the color contour of the part, but with the contours scaled to match those of the previous study: **Results** → **Displacements**, then **Chart Options** → **Display Options** → **Defined**

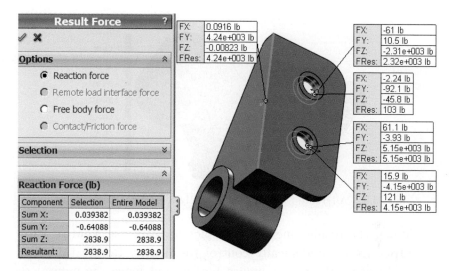

Result Force

Options

- ● Reaction force
- ○ Remote load interface force
- ○ Free body force
- ○ Contact/Friction force

Selection

Reaction Force (lb)

Component	Selection	Entire Model
Sum X:	0.039382	0.039382
Sum Y:	-0.64088	-0.64088
Sum Z:	2838.9	2838.9
Resultant:	2838.9	2838.9

FX:	0.0916 lb
FY:	4.24e+003 lb
FZ:	-0.00823 lb
FRes:	4.24e+003 lb

FX:	-61 lb
FY:	10.5 lb
FZ:	-2.31e+003 lb
FRes:	2.32e+003 lb

FX:	-2.24 lb
FY:	-92.1 lb
FZ:	-45.8 lb
FRes:	103 lb

FX:	61.1 lb
FY:	-3.93 lb
FZ:	5.15e+003 lb
FRes:	5.15e+003 lb

FX:	15.9 lb
FY:	-4.15e+003 lb
FZ:	121 lb
FRes:	4.15e+003 lb

Fig. 4.7.　Reaction forces from the washers, bolts, and back edge.

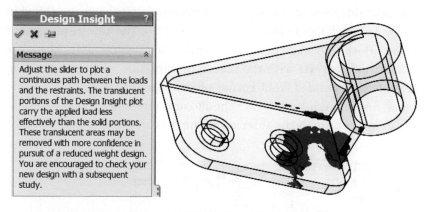

Design Insight

Message

Adjust the slider to plot a continuous path between the loads and the restraints. The translucent portions of the Design Insight plot carry the applied load less effectively than the solid portions. These translucent areas may be removed with more confidence in pursuit of a reduced weight design. You are encouraged to check your new design with a subsequent study.

Fig. 4.8.　Major material "load path" through the part.

and set the prior maximum value. Figure 4.9 (left) shows the new displacement distribution which is less that half as large as before, and includes smaller regions of bending.

To add a more informative displacement vector plot:

(1) Right click in the graphics area. Pick **Edit Definition** → **Advanced Options** → **Show as vector plots**.

Fig. 4.9.   Alternate restraints significantly reduce peak deflections.

(2) When the displacement vector plot appears use **Vector Plot Options** to dynamically control the clarity. Rotate this view (hold down the center mouse button and move the mouse) to find the most informative orientation.

The right side view of the original displacements is shown amplified in Figure 4.10 (left) and compared on the right to the new displacement values. Both are shown at the same scale, along with the gray undeformed part. As with the first approximation, the main bending region is the rounded front corner of the base plate. The mesh there is now fine enough to describe the flexural stress and shear stress as they change through the thickness. The default contour plot style for

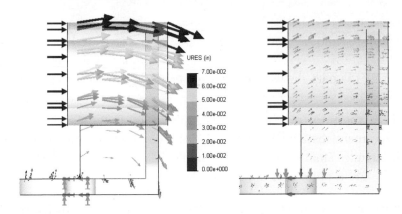

Fig. 4.10.   Original (left) and revised displacements, to the same scale.

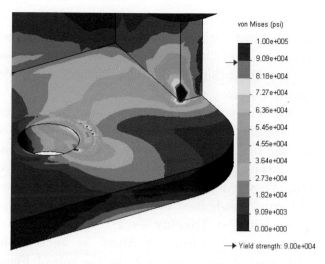

Fig. 4.11. Distortional energy failure criterion on the base top surface.

viewing the displacements is a continuous color variation. These are pretty and should be used at some point in a written stress analysis report, but they can hide some useful engineering checks.

The effective stress distribution on the top of the base is shown in Figure 4.11. There two different contour options are illustrated. The line contour form tends to be more useful if preliminary reports are being prepared on black and white printers. The material yield stress was 90,000 psi, so every region colored in red is above yield. In that figure, the **Edit Definition** feature was used to manually select the discrete color or line contours. The **Chart Options** feature was used to specify the stress range and the number of discrete colors. Rotating the part shows that there are other regions of the part above the yield stress. On the bottom corner of the base the assumed line support has also caused high local stresses. In that figure you should also note that the bottom edge of one bolt bearing surface shows a small region of yielding. That should be examined in more detail and other views.

Line or point restraints are not likely to exist in real part components. You could revise that region by putting in a split line to create a narrow triangular support area. That would be more accurate, if the resulting part stresses there are compressive. Otherwise, you would need to use a contact surface. Using contact surfaces where gaps can develop requires a much more time consuming iterative solution. However, the important thing is to attempt an accurate model of the part response, not an easy model.

The extreme values of the stress tensor at any point are known as the principal stresses (eigenvalues of the stress tensor). For 3D part, there are three normal principal stresses and a maximum shear stress. The principal normal stress components have both a magnitude and direction. They can be represented as directed line segments with two end arrow heads used to indicate tension or compression. Employ **Stress Plot → Display → P1 Maximum principal stress → Advanced Options → Show as vector plot**. The maximum principal stress vector plot, at the top of Figure 4.12, shows the highest tension stress (and is a failure criterion for some materials). Here, it occurs in the yielding region where the base joins the vertical tube support. Those high stress concentrations could (and should) be reduced by adding fillets along the associated edges along the reentrant corners. The display is from **Advanced Options → Nodal values.**

The SW Simulation system allows the user to select from a short list of the most studied material failure theories. It can display a plot

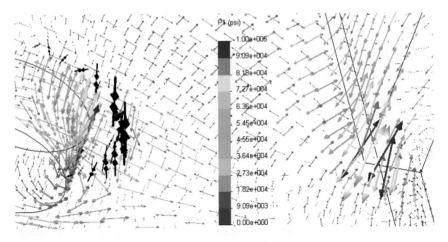

Fig. 4.12. Maximum principal stress vectors.

Fig. 4.13.   Material failure is predicted for the current study.

of a non-dimensional number representing the value of the theory at every point on the part. A value less than unity means that the selected theory predicts material failure. Many engineers refer to that value as the factor of safety while others refer to it simply as the failure theory factor that is one of many terms, listed in Table 3.1. To see this type of plot: right click on **Results → Define Factor Of Safety Plot** to open the **Factor of Safety panel** and in this case select the von Mises criterion, OK. All the factors in that table need to be unity or greater. The resulting plot in Figure 4.13 shows a minimum material failure theory value of about 0.6. Thus, this design is currently a failure.

4.10. Validate the Design Revision

The reaction results, in Figure 4.7, suggest an assumption error in this study. In particular, it is important to check the signs of the reactions to verify that they act as you expect. The negative sign of the left bolt reaction in the z-direction was unexpected. To better illustrate the concern, and to show another post-processing tool, consider a free body diagram of the regions governing reactions in the z-direction: right click **Results → List Force Results → Options → Free body force**. Under **Selection** set the units and pick the pressure face and the two bolt bearing faces, Figure 4.14.

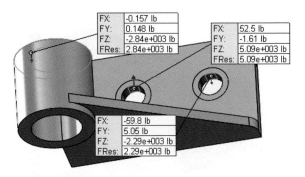

Fig. 4.14.　Free body diagram shows error in the support assumption.

The pressure resultant force (about 2,840 lb) and the left bolt force (about 2,290 lb) both act in the negative z-direction. They are resisted by the positive z-direction force (about 5,090 lb). The forces in the z-direction are in equilibrium (within 0.08% round off error). Consider moment equilibrium about the y-axis, at the center of the right bolt. The x-lever arm to the pressure resultant center is 53 mm, while the lever arm to the left bolt is 65 mm. Thus, y-moment equilibrium, $(2.84e3*55 - 2.29e3*65 \approx 0)$ is also satisfied (to 1% round off).

The concern with the reaction results is that the left bolt was assumed to be pushing on the front bearing area. Instead, it is pulling on that area, which is not possible (a gap would open). What is likely is that the bolt will simply push on the corresponding area at the back of the hole with the same reaction force. To allow for that support mode, the first split line should have been made on the back of the bolt hole. That can be easily fixed. An alternative would have been to build an assembly to include part of the bolts and conduct an iterative contact/gap analysis. That would have take much longer (and sometimes fails to converge), but that is often not needed in a preliminary study. Another alternative is to take advantage of the standard bolt connector application available in SW Simulation. The question is, is that support (fixture) change likely to invalidate the above study results? Looking at the load path sketch of Figure 4.8 and the stresses in Figure 4.11 it is unlikely. However, the next study refinement of this part should utilize this new information.

The main goal of this study was to illustrate the usefulness of split surfaces, the need for engineering judgment in making restraint assumptions, and typical ways to examine the results of a study. The locations and type of restraints are usually the least well know aspect of a part analysis or design. Load conditions are probably the next least reliable information. Use friendly software to investigate various combinations of loads and restraints to get the safest results. Also check independent solutions when possible.

4.11. Other Aspects of Mesh Generation

SW Simulation has a very powerful solid mesh generator for tetrahedral elements. Almost every analysis requires the engineer to employ judgment on where to apply controls to the mesh. Usually the mesh needs to be refined around restraint regions, load regions, and reentrant regions of the solids. Figure 4.15 illustrates that you can control element sizes on faces, edges, and vertices. The earlier example showed that you should plan ahead while building your solid to allow for expected mesh control needs and insert split lines into your solid to allow for additional entities to be selected on your model. Mesh control lets you specify the desired element size, and the rate at which adjacent elements increase in size.

Any sharp (no fillet) reentrant corner in theory causes infinite radial derivatives at the corner in heat transfer and stress analysis studies. A fillet removes the infinite stresses, but interior fillets

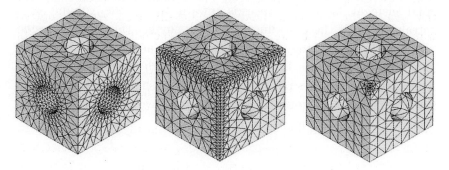

Fig. 4.15. Options for local mesh control: surface, curve, or vertex.

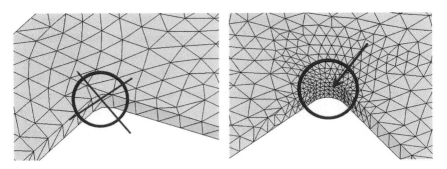

Fig. 4.16.   Avoid large elements along arcs and through the thickness.

usually need a reasonable amount of mesh refinement, as illustrated in Figure 4.16. Exterior fillets are less important in getting accurate results and are sometimes suppressed in finite element studies. Split lines are optional for some solids but are often mandatory to properly join and align shell meshes. This is also important when connecting shells to solids.

Bad solid modeling practices are probably the most common cause of failure of the mesh generation. The mesh generator begins with the edges of a solid. It divides them into segments corresponding to the requested element sizes. However, if the mesh generator encounters a line segment that is smaller than the requested element size, then it must decrease the minimum element size the match the short line. After processing the boundary lines, the generator proceeds to the part surfaces. Then it fills those faces with triangles of the minimum to the requested size. From the triangles (initial solid element faces) it proceeds inward to fill the volume with tetrahedrons. If the edge lines are much smaller that the requested element size, or if they join at very small angles then the mesh generator tries to use tiny elements to match the poor local part geometry. If it is possible to do that the mesh is still likely to fail due to insufficient computer memory. Part flaws are often too small to see without zooming in on the part.

5

General Solid Stress Analysis

5.1. Introduction

Every part, at some level, can be thought of as a 3D solid. That is the default analysis mode of SW Simulation. You usually start every study by building a solid, even if you in turn model it as a shell or frame. Therefore, you need to learn the numerous options that are available to support such solid stress studies. To validate the results of a 3D solid study you often need to use an analytic approximation or an FEA beam, frame or shell model. For the proper assumptions, those lower dimensional studies can be quite accurate and are almost always much less demanding of computer resources. You will find that you never have large enough computer resources and you will have to learn how to use symmetry, anti-symmetry, beams, frames, shells, and trusses to reduce some problems to a size that can be solved with your available resources. At other times you will use those procedures as a way to independently validate a more complex study.

5.2. Flexural Analysis of a Zee-Section Beam

In this study you will validate your understanding of the use of SW Simulation by solving a cantilever beam and comparing the FEA results to that predicted by mechanics of materials theory. The constant cross-section is a zee-shape in the x–y plane as seen in Figure 5.1. It extends in the z-direction for a length of $L = 500$ mm. The thickness of the section is $t = 5$ mm, each flange has a length of $a = 20$ mm, and the web has a depth of $h = 2a = 40$ mm. At the

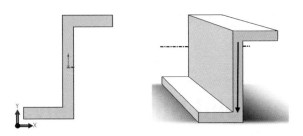

Fig. 5.1.   A zee-section straight beam solid.

free end it is loaded by a distributed force parallel to the y-axis (i.e., vertical).

Before you start an FEA study you should try to get a reasonable approximation of the stresses and deflections to be obtained. This can be an analytic equation for a similar support and loading case, an FEA beam model compared to a continuum solid model, or a one or two element model that can be solved analytically, etc.

The cantilever is horizontal and has a vertical load of $P = 500\,\text{N}$. Therefore, it causes a bending moment, about the x-axis of $M = P(L - z)$, where z is the distance from the support. Such a loading causes a linear flexural stress (σ_z) that varies linearly through the depth. For symmetric sections (*only*) that stress is zero at the neutral axis (here parallel to the x-axis at the section centroid) and has a maximum tension along the top edge, and a compression along the bottom edge (parallel to x). The load P causes a varying moment and a shear force. The corresponding transverse shear stress (τ) varies parabolically through the depth and has its maximum at the neutral axis. Those (symmetric) stress behaviors are sketched with the section in Figure 5.2. The flexural and shear stress equations are $\sigma_z = My/I_x$ and $\tau = PQ/tI_x$ where I_x is the second moment of inertia of the section and Q is the first moment of the section at a distance, y, from the neutral axis. For this section $I_x = 2ta^3/3$. The maximum tension will occur at $y = a + t/2$, while compression occurs at $-y$. Likewise, the end deflection of the beam in the vertical (y) direction will be $U_y = PL^3/3EI_x$. With that (symmetric) beam review and its predictions you can now proceed with the FEA study.

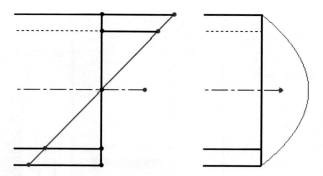

Fig. 5.2. Flexural (left) and shear stresses from thin beam theory.

Select the SW Simulation icon:

(1) Right click on the **Simulation → New Study**. Select **Static** and enter the new **Study Name** (Zee_beam).
(2) Right click on **Solids → Apply/Edit Material**. In the **Material panel** pick **From library files**, and select SI units. Select **Copper Alloys → Brass** and review the properties (and significant figures).
(3) Right click on **Fixtures** to open the **Fixture panel**. Select **Fixed Geometry** and select the wall end of the beam. Note that while rotational fixture icons appear, they are not present in solid elements.

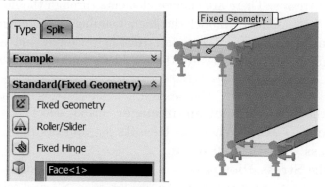

(4) Right click on **External Loads → Force**. Select the beam free end **face**, a vertical **edge** for the direction, and set the **value** at 500 N.

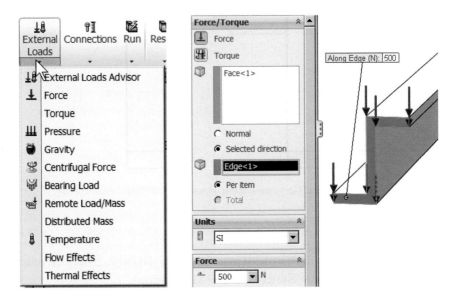

(5) Next create a default mesh, right click on **Mesh** → **Create Mesh** → **OK**.

Since local top or bottom flange bending is not expected to be high, the default mesh with only one quadratic solid through the thickness should be acceptable. Otherwise you should have at least three elements through the thickness in a region of expected local bending stresses. There are enough elements through the height and length of the solid to model the high bending expected near the **immovable** (cantilever) restraint. Having reviewed and accepted a default mesh you execute the problem and recover selected results:

(1) In the **SW Simulation manager** menu select the **Study Name** → **Run**. When the **Results** list appears right click on **Stress** → **Edit Definition**.

(2) In the **Stress Plot** panel select **SZ: Z normal stress** as the component and **Fringe** as the display type. SZ was selected as the first display since it is the normal stress component parallel to the beam axis that you would validate with beam theory (Figure 5.3).

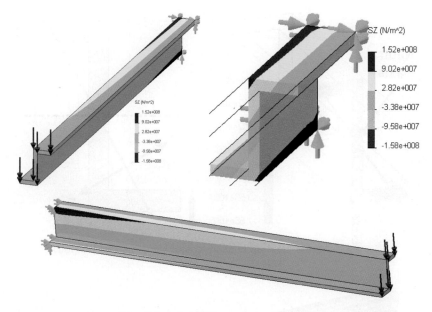

Fig. 5.3.   Axial flexural stress levels.

(3) Optionally control the stress display by right clicking in the
    graphics area **Settings → Settings panel → Discrete**
    fringe options, and to better see the stress differences: Right
    click in graphics area **Chart Options → Chart Options
    panel → 5** color levels.

The resulting stress contours have a maximum value of about
152 MPa, in both tension and compression. But, the stress contour
spatial distributions are not what you would expect from symmetric
beam theory. That theory predicts the flexural stress contours on the
top and bottom to be parallel to the restraint wall (perpendicular to
the beam axis). Yet the actual stress contours are almost parallel
to the beam axis. In other words, symmetric beam theory predicts
a neutral axis (NA), at the beam half depth and parallel to the
flange.

That is, the NA would be expected to be parallel to the global
x-axis. Instead of zero normal stresses there, they are zero along an
inclined line rotated about 55 degrees w.r.t. the x-axis. The actual

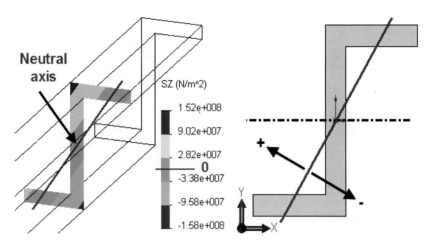

Fig. 5.4.   Actual bending neutral axis of the cross-section (red).

Fig. 5.5.   von Mises effective stress distribution.

NA is highlighted in Figure 5.4. Points above the NA are in tension here and those below are in compression. Figure 5.5 shows a similar distribution for the von Mises stress. Since the stress distribution is quite different from symmetric beam theory you should also look at the deflections in detail. Symmetric beam theory says that the deflection is a maximum at the free end and lies in the z–y (side) plane and there is no deflection in the z–x (top) plane.

However, Figure 5.6 shows that there are significant displacements out of the plane of the beam web and resultant loads. The graphs in Figure 5.7 verify that the horizontal (top view) deflections are larger than the (side view) vertical deflections. Therefore, there was something wrong with the 1D mechanics of materials concept

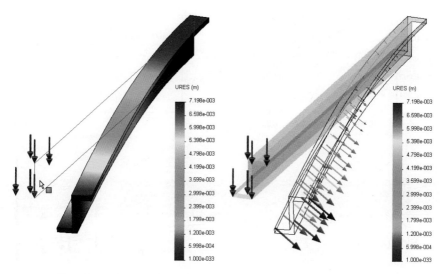

Fig. 5.6.   Out of plane (lateral) displacements of zee member.

that was selected to predict the results of the solid finite element study (or you made an error).

The simplified predicted deflection of $U_y = 0.0078$ m is almost twice as large as $U_y = 0.0042$ m found in Figure 5.7. That figure also shows a value of $U_x = 0.0059$ m while the simplified theory predicts zero. Likewise, the simplified normal stress estimate, at a constant axial z position, predicts a constant stress along the top and bottom edges of the beam, but *that was not observed in the solid solution. Of course the FE results and the validation predictions do not* agree! The simplified 1D beam mechanics relations are only valid for straight symmetric sections. Usually those sections have two planes of symmetry. But they must have at least one symmetry plane, so as to make the product of inertia vanish ($I_{xy} \equiv 0$). The current section does not have a single symmetry plane. Its geometric inertias are $I_x = 2/3\,ta^3$, $I_y = 3/8\,ta^3$, $I_{xy} = -ta^3$, and its cross-sectional area is $A = 4\,ta$. The 1D unsymmetrical beam theory predicts that the NA axis will rotate from the x-axis by an angle of $\alpha = \tan^{-1}(-I_{xy}/I_x) = 56.3$ degrees, which seems to agree with Figure 5.4. Any time $I_{xy} \neq 0$, one must employ non-symmetric beam theory for bending.

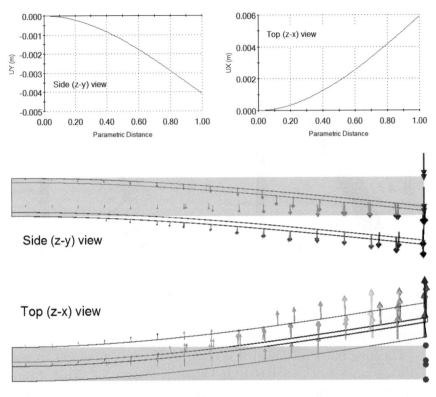

Fig. 5.7.   True shape displacement data for zee member.

The general beam theory [10] states that the cross-sectional normal stress varies with both x- and y-positions in the cross-section according to the equation:

$$\sigma_z = \frac{(M_x I_y - M_y I_{xy})y - (M_y I_x - M_x I_{xy})x}{I_x I_y - I_{xy}^2}$$

but since in this case $M_y = 0$ so the bending stress is

$$\sigma_z = \frac{M_x(I_y y - I_{xy} x)}{I_x I_y - I_{xy}^2} = \frac{PL(6y - 9x)}{7ta^3},$$

which reduces to the original stress estimate only for a symmetric beam, $I_{xy} = 0$. On the topmost horizontal line of the flange ($y = a + t/2$) the above stress is estimated to vary from a corner tension of about $PL/7ta^2$ to $3/2$ of that value in compression. The more general non-symmetric beam 1D displacement predictions are

$$U_y = \frac{PI_y(L-z)^3}{6E(I_xI_y - I_{xy}^2)} + c_0z + c_1 = \frac{P[(L-z)^3 + 3zL^2 - L^3]}{7Eta^3},$$

$$U_x = \frac{PI_{xy}(L-z)^3}{6E(I_xI_y - I_{xy}^2)} + c_2z + c_3 = \frac{-3U_y}{2},$$

which yield maximum values of

$$U_{y\,max} = \frac{-2PL^3}{7Eta^3}, \quad U_{x\,max} = \frac{-3U_{y\,max}}{2}, \quad U_{max} = \frac{0.515PL^3}{Eta^3}.$$

For the given dimensions the above estimates reduce to $\sigma_{z\,corner} = 17.9e7$ MPa at the restraint wall and $U_y = -0.0045$, $U_x = 0.0067$, and $U_{max} = 0.0081$ meters at the free end, respectively. The new validation estimates agree with the solid study results reasonably well. The results do suggest using a finer mesh in the corners near the restraint wall.

5.3. Ram Block Stress Analysis

A pressure container is formed from a brick of corrosion resistant steel. The block contains a center large cylindrical hole. Orthogonal to that is a second oval hole coming from the side and also going all the way through the block. The block is 24.75 inches square and 42 inches long. Its central cylindrical hole is 18.75 inches in diameter. The oval intersecting passage matches the inner diameter, but has two 5 inch radius semi-circular ends on a rectangular center, as seen in Figure 5.8.

The container is subjected to a (self-equilibrating) constant pressure of 3,000 psi. Due to the symmetry in the geometry, materials, and pressure you can utilize a one-eighth symmetry study for the dimensions of the corner component given in Figure 5.9. The main purpose of this example is to show how to do an analysis and how to take advantage of the many graphical features that can be selected to enhance your understanding of a general 3D component. They also help in documenting your written report.

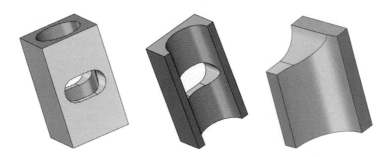

Fig. 5.8.   The ram pressure block, half and one-eighth symmetric regions.

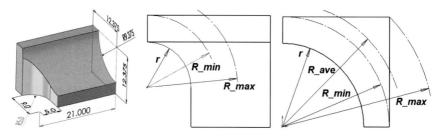

Fig. 5.9.   Corner geometry with dimensions, and approximate cylinders.

It is desirable to try to estimate the stresses and deflections to be encountered. You will not find a handbook solution for a rectangular block with a pressurized cylindrical hole. However, you can find stresses and deflections for a thick walled cylinder with internal pressure (and free or fixed end walls). The axial stress is zero; the maximum radial stress is at the inner radius and is in compression. The hoop stress is tension and also maximum at the inner wall. Thus, you can find the principal stresses, and von Mises stress, at the inner radius. That wall is cut through by the oval channel. That cutout will cause a "stress concentration" where it cuts the inner wall (and elsewhere). There are handbook solutions for very similar elliptical openings. Combining such solutions may get us close to the stresses and deflections.

To prepare a validation estimate for the maximum deflection you need to consider a range for the effective outer radius of the cylinder, R (right of Figure 5.9). For example, you usually want to consider an average value, R_{ave}, that has the same cross-sectional area. You could get an upper bound estimate by using the minimum wall thickness. The maximum radial displacement is tabulated as $\delta_{max} = 2PRr^2/[E(R^2 - r^2)]$. For and average thickness $\delta_{max} = 3.7e\text{-}3$ inches.

The maximum axial, hoop, and radial (principle) stresses of a cylinder from mechanics of materials are axial $\sigma_1 = 0$, circumferential $\sigma_2 = P(R^2 + r^2)/(R^2 - r^2)$, and radial $\sigma_3 = -P$, respectively. Here, P is the internal pressure, r is the inner radius, and R the outer radius. Using the minimum wall thickness, the estimated stresses give a maximum von Mises value of about 12.8 ksi.

The stress concentration factor, K, for an elliptical hole in a plane stress surface is known for a few ratios of σ_2 and σ_3. It depends on the ratio of the major and minor diameters of the ellipse. Here that diameter ratio is about $a/b = 18/9 = 2$. The closest tabulated case is for equal magnitude tension and compression in the material around the hole. Then $K = 2(1 + a/b) = 6$, but the higher compression stress would possibly lower it to as low as 4. Thus, estimate the maximum stress as $\sigma_{max} = 53$ ksi, which is above the yield stress of 41 ksi. Begin the study by selecting your preferred units:

(1) **Tools → Options → Document Properties → Units.** Check **IPS** (inch, pound, second) as your **Units, OK.**
(2) Select the **Simulation → New Study** to open the Manager menu.
(3) Enter PV_stress as the **Study name**, and **static** as the **Analysis type**.

The material is expected to operate at a temperature of about 400 F and contain a corrosive fluid. Either of those two conditions makes it unlikely that the standard library of materials will contain the alloy you need, C276 steel. Define the material properties:

(1) Right click on the solid **Part name → Apply/Edit Material** in the **Manager menu.**
(2) When the **Material panel** opens, check in **SolidWorks Materials.**

(3) Click on Steel to expand the list of alloys, and check it for C276.

(4) Since that is not found you must create a custom material.

The process for creating custom material changed with the release of SolidWorks 2010. You must first open the material library, copy a standard material, edit its properties, and save it with a new name. The steps are:

Right-click on the **Part name** and select **Edit Material**.

(1) In the **Material panel**, select the material on which to base the custom material. Right-click on that material name. Its properties appear in the Properties tab. Right click and select **Copy** to copy the material to your paste buffer.

(2) In the **Material panel**, expand the **Custom Materials tree**. **Paste** the property there.

(3) In the **Properties Tab**, first change the material name in the Name box. Next enter new property values. The required property names for a given type of study appear in red.

(4) Pick **Save** to add the new material to the library.

(5) Click **Apply** to apply the new material to the current part, and then pick **Close**. Otherwise, select Close and confirm that you wish to save the new property set.

For the current study, type C276_steel_400F as the name. Enter all the required (red) mechanical material properties ($E = 29.8$ ksi, $\nu = 0.31$) as well as the thermal properties ($k = 1.74e\text{-}4$ BTU/in-s-F, $\alpha = 7.1e\text{-}6$). The thermal properties are not required here, but will be used to compute a thermal study or a thermal stress study later. To find properties at a specific temperature you may have to search the Web or a handbook. Check for units consistent with the table.

Define displacement restraints that reflect the chosen symmetry, and eliminate the six rigid body motions:

(1) In the **Manager menu**, select **Fixtures** → **Advanced** → **Symmetry** to open the first fixtures panel.

(2) Pick **Symmetry** as the **Type** and select the two rectangular vertical faces as the **Selected Entity**, along with the horizontal

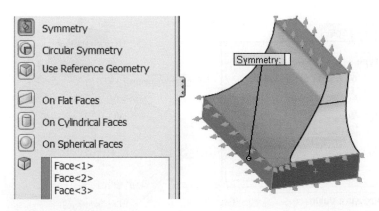

Fig. 5.10. Restrain the three symmetry planes and all the RBM.

symmetry plane at the center of the ram opening. They prevent all six rigid body motions.

(3) Visually check the prevented motions seen in Figure 5.10. When satisfied pick **OK** (green check mark).

The entire internal surface will be subjected to the pressure:

(1) In the **Manager menu**, select **External Loads** → **Pressure** to open the **Pressure panel**. Set the **Pressure Type** to be **normal to selected face** (in Figure 5.11).

(2) Select each of the three internal curved surfaces. Under **Pressure Value** set English **Units** and type in 3,000 psi as the **Value**.

(3) Click **Preview** to visually check the (red) pressure arrows around the perimeter of each surface. (Changing the sign of the value reverses the arrows.) When satisfied pick **OK**.

Note that a symmetric thermal loading could be included in the study, but gravity directed along the cylinder axis, for example, would require using a half part model and more computer resources. Since this is the first study, you probably will have to edit the part geometry and repeat this study. Avoid larger models until initial refinements are completed.

Next the mesh needs to be created. For an initial study you might get by with a default mesh size. However, an accurate mesh almost always requires engineering judgment to control the element sizes

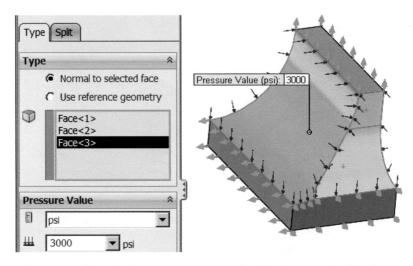

Fig. 5.11.   Specify the internal pressure distribution on curved surfaces.

(or a high quality automatic error estimator). The three symmetry planes contain the smallest wall thicknesses, so as a first cut the mesh should be make smaller there:

(1) In the **Manager menu**, right click on **Mesh** → **Apply Control** to open a **Mesh Control panel**.
(2) Pick each of the three symmetry planes to go in **Selected Entities**.
(3) Note the suggested default **element size** and reduce it to about 0.5 inches (for these 3 inch thick walls). Accept the default **transition rate** of 1.5 for adjacent element sizes (Figure 5.12).
(4) Pick **Preview** to verify the controlled entities (not shown), **OK**.

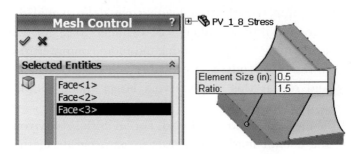

Fig. 5.12.   Initial surface mesh size control.

The curved intersection line of the two passages might also be of concern. If so, you could have a second **Apply Control**, select those curve segments and specify the desired element sizes on that edge. When you are finished using your engineering judgment for planning the mesh, via **Apply Control**, then: In the **Manager menu**, right click on **Mesh → Create** to open a **Mesh panel**. There, move the **slider bar** to make the average volume elements smaller, **OK**. The initial mesh, looking toward the pressure surfaces, is in Figure 5.13.

The mesh has three unknown displacements per node. To execute the static solution for those displacement components and to recovery the stresses, right click on **Study name → Run**. Passing windows will keep you posted on the number of equations being solved and the status of the displacement solution process and post-processing. You should get a notice that the analysis was completed (not a failed message). Then you have access to the various Simulation report and plot options needed to review the first analysis.

The default post-processing plot is a smoothly filled (Gouraud) contour display of the requested variable. If you do not have a

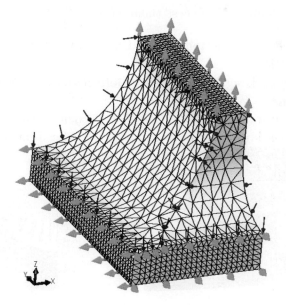

Fig. 5.13.   Initial solid element mesh.

color printer and/or if you want a somewhat finer description you may want to change the default plot styles. After a default plot appears:

(1) Right click in the graphics window and select **Settings** → **Fringe Options**. Change from the default **Continuous** Gouraud filled image to a **Discrete** or **Line** contour option.
(2) Select **Superimpose model** if you wish to also see the undeformed shape in the same display.

To review the deformed shape magnitude:

(1) Double click **Plot 1** under **Deformation** in the **Manager menu**, Figure 5.14. Double click **Plot 1** under **Displacement** in the **Manager menu**. Rotate the view. Right click in the graphics area and select **Color Bar** to control the contour ranges.
(2) Right click in the graphics area, **Edit Definition** → **Settings** → **Include undeformed part**.

Displacements are vector quantities; therefore consider a vector plot first, as in Figure 5.15. Access them with:

(1) Right click in the graphics area, **Edit Definitions** → **Displacement Plot**.
(2) Double click again and **Edit Definitions** → **Vector Plot Options**.

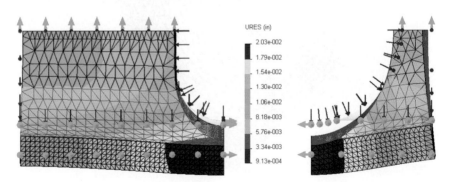

Fig. 5.14.   Scaled deformed mesh shape.

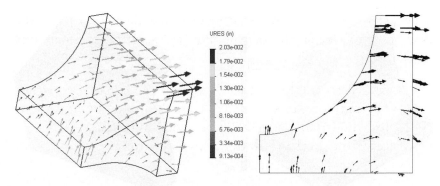

Fig. 5.15.   Two views of ram block resultant displacement vectors.

The predicted validation result was based on a thick walled cylinder. If the solid results were in close agreement then the displacement vectors, as seen from the end (Figure 5.15 right) would have mainly been in the radial direction, but they are not. To try to compare the solid results to a cylinder, the radial displacement component with respect to the long cylinder axis is displayed in Figure 5.16. The average radial displacement is reasonably close (less than a factor of two).

To view the deformed shape in a radial direction use **Plot** → **Advanced** → **Axis** and pick local **Coordinate System 1**. The computed outer deflection (Figure 5.14) was about 2e-2 inches, which is near the radial estimated range of 1.6e-2 to 2.6e-2 inches.

Next check the stress component levels by double clicking on **Stress** → **Plot** icon. There are many types of stress evaluations

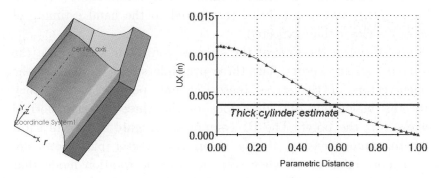

Fig. 5.16.   Radial displacement (coordinate system 1) at free end arc.

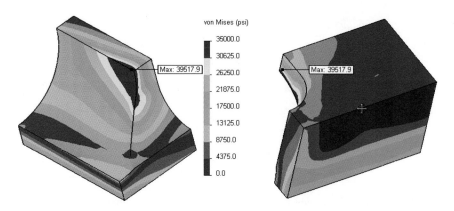

Fig. 5.17.   Peak von Mises stress levels.

available. The default one is the scalar von Mises (or Effective) stress. It is actually not a stress but a failure criterion based on distortional energy, for ductile materials, that has the units of stress. Since you picked a ductile material, the von Mises value should be examined and compared to the material yield stress. Figure 5.17 shows the effective stress from different points of view. The maximum effective stress is seen to be about 40 ksi which is barely below the material yield stress. The validation estimate of the von Mises stress was 53 ksi, which is within about 25%. It is almost always difficult to get reasonable stress validations without a second different type of FEA.

The computed von Mises value (Figure 5.17) of 38 ksi is close to that value and the hand solution. The computed σ_1 (P1 of Figure 5.18) was about 40 ksi (compared to the hand estimate of 53 ksi) is closer than expected.

To visualize the flow of the stresses you often wish to see the principle stress vectors. The three principle stress vectors are the eigen-vectors of the stress tensor at any point. That is, they are the magnitude and directions of the three extreme normal stresses at the point. The P1 component (Figure 5.18) will show the maximum tension, if any is present. Vector plot views are even more informative when seen in dynamic rotation mode that is available with (highly recommended) 3D mouse hardware.

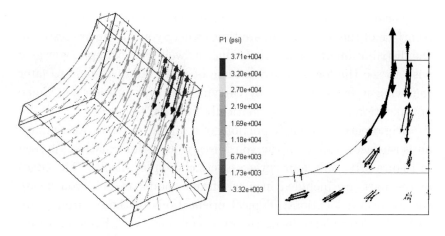

Fig. 5.18.   Two views of the maximum tension stress, P1.

Isosurface displays are surfaces of constant values within the solid elements. They are similar to contours, but are computationally intensive to generate and rotate. It is best to display them with a small number of surfaces. To see isosurfaces of an item:

(1) Right click in the graphics area and select **Edit Definition**. For a stress item the **Stress Plot** → **Display panel** will open (Figure 5.19).
(2) Under **Plot type** check Iso, use a discrete **Fringe type** and 4 as the **No. of surfaces**.
(3) Select the desired **Component** and nodal values **Result Type**, **OK**.

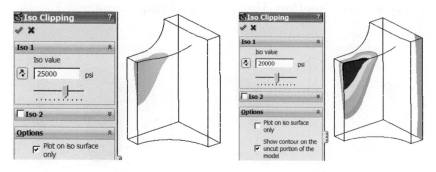

Fig. 5.19.   Two options for selecting isosurfaces for stress components.

The extreme values of a stress component in a solid will occur on a surface of the solid. There are times where you will be interested in the distribution of the interior values as well. For example, you may wish to see the low stressed volume of material as a guide to later removing it from the part by carrying out and extrude cut. Some optimization codes can automatically function that way.

To potentially identify portions of a ductile material for removal an isosurfaces display of the von Mises value is informative. As seen in Figure 5.19, only four surfaces were chosen for display. The edges of the part are included in a wireframe model to help you locate the isosurface locations. Typical orientations of those surfaces are shown in Figure 5.20, using the same contour levels. Even with that small number of surfaces you can see that the relatively low stressed volume is large. Such images are slow to change, in view rotations, even with 3D mouse hardware.

Contours shown on cutting planes can also be useful in seeing the internal distribution of an item. If you utilize only one cutting plane they are basically section views with contours added to them. Then you have the choice of also seeing contours on either the surface in front of, or behind the flat cut plane. SW Simulation offers the ability to have multiple flat cut planes, with contours displayed, at the same time. Of course, you can control both their orientation and location. To activate such a plot for an item:

(1) Right click in the graphics area and select **Edit Definition**.
(2) For a stress item the **Stress Plot → Iso Clipping**. Under **Plot type** check **Section**. Use a discrete **Fringe type** and 1 as the

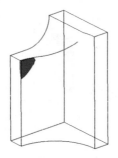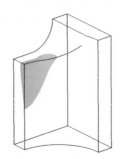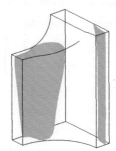

Fig. 5.20.   von Mises isosurfaces at 35, 25, and 15 ksi.

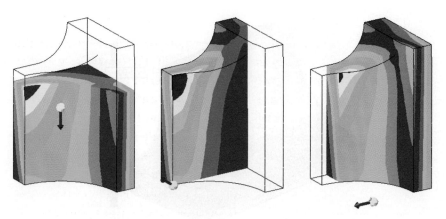

Fig. 5.21.   von Mises contour levels for standard clipping planes.

No. of sections. Select the desired **Component** and **Result Type**, **OK**.

(3) Right click in the graphics area, pick **Clipping** to open a **Section Clipping panel**.

(4) There use the **slider bars** to set the **(X, Y, Z)** components of the unit vector normal to the cutting plane.

(5) Use the **Position slider** to dynamically position the plane in the part.

Using the standard cutting planes, the von Mises value was chosen for display and the resulting images are in Figure 5.21. You can also select cutting surfaces that are cylindrical or spherical.

A useful feature of SW Simulation is to allow the engineer to select a material failure criterion and the plot the value of the material factor of safety based on that choice. The FOS should be greater than 1, and typically less than 4. Such a plot is obtained with:

(1) Double click on **Design Check** → **Plot 1** in the **Manager menu**.

(2) Select the proper **failure criterion** for the material you have used.

(3) Execute the plot for the result in Figure 5.22. The minimum safety values found were 1.02 for maximum shear stress, 1.04 for von Mises, and 1.11 for maximum normal stress. All are too low.

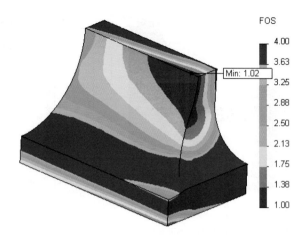

Fig. 5.22.   Material factor of safety view based on maximum shear stress.

From a series of isosurfaces of various failure modes, like Figure 5.21, you may be able to estimate the major load paths. The **Results → Define Design Insight Plot** (pick the wireframe display mode first) provides that information via a slider scale.

The result here shows extremely low (≈1) values for the material FOS. You must either drastically reduce the design pressure, which is not likely to occur, or significantly increase the wall thickness. You could easily place a small fillet along the high stress hole intersection in hopes of reducing peak stress levels.

A less easy task is to envision some way to pre-stress the current part so it would develop initial compression in the high tension regions seen here. A completely different approach would be to find a near optimal shape using finite element topological optimization. Then you would build a solid geometry that is very close to the computed optimal shape (which is usually quite irregular). The thermal analysis of this part and the resulting thermal stresses are conserved in later sections.

6

Plane Stress Analysis

6.1. Introduction

Generally you will be forced to utilize the solid elements in SW Simulation due to a complicated solid geometry. To learn how to utilize local mesh control for the elements it is useful to review some two-dimensional (2D) problems employing the membrane triangular elements. Historically, 2D analytic applications were developed to represent, or bound, some classic solid objects. Those special cases include plane stress analysis, plane strain analysis, axisymmetric analysis, flat plate analysis, and general shell analysis. After completing the following 2D approximation you should go back and solve the much larger 3D version of the problem and verify that you get essentially the same results for both the stresses and deflections.

Plane stress analysis is the 2D stress state that is usually covered in undergraduate courses on mechanics of materials. It is based on a thin flat object that is loaded, and supported in a single flat plane. The stresses normal to the plane are zero (but not the strain). There are two normal stresses and one shear stress component at each point (σ_x, σ_y, and τ). The displacement vector has two translational components (u_x, and u_y). Therefore, any load (point, line, or area) has two corresponding components.

The SW Simulation "shell" elements can be used for plane stress analysis. However, only their in-plane, or "membrane", behavior is utilized. That means that only the elements in-plane displacements are active. The general shell in-plane rotation vectors are not used for plane stress studies and should be restrained. To create such a study you need to construct the 2D shape either as a sheet metal part, a planar surface (**Insert → Surface → Planar surface**), or extrude it as a solid part with a constant thickness that is small compared to the other two dimensions of the part. A solid is later converted to a shell model by creating a mid-surface or an offset surface within the solid.

Before solid elements became easy to generate it was not unusual to model some shapes as 2.5D. That is, they were plane stress in nature but had regions of different constant thickness. This concept can be useful in validating the results of a solid study if you have no analytic approximation to use. Since the mid-surface shells extract their thickness automatically from the solid body you should use a mid-plane extrude when you are building such a part.

One use of a plane stress model here is to illustrate the number of elements that are needed through the depth of a region, which is mainly in a state of bending, in order to capture a good approximation of the flexural stresses. Elementary beam theory and 2D elasticity theory both show that the longitudinal normal stress (σ_x) varies linearly through the depth. For pure bending it is tension at one depth extreme, compression at the other, and zero at the center of its depth (also known as the neutral-axis). When the bending is due, in part, to a transverse force then the shear stress (τ) is maximum at the neutral axis and zero at the top and bottom fibers. For a rectangular cross-section the shear stress varies parabolically through the depth. Since the element stresses are discontinuous at their interfaces, you will need at least three of the quadratic (6 node) membrane triangles, or about five of the linear (3 node) membrane triangles to get a reasonable spatial approximation of the parabolic shear stress. This concept should guide you in applying mesh control through the depth of a region you expect, or find, to be in a state of bending.

6.2. Rectangular Beam Segment

A simple rectangular beam plane stress analysis will be illustrated here. Consider a beam of rectangular cross-section with a thickness of $t = 2\,\text{cm}$, a depth of $h = 10\,\text{cm}$, and a length of $L = 100\,\text{cm}$. Let

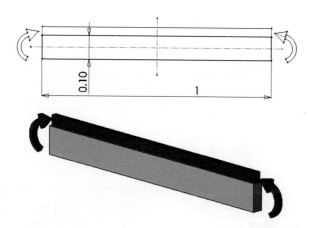

Fig. 6.1. A simply supported beam with line load and end moments.

a uniformly distributed downward vertical load of $w = 100\,\mathrm{N/cm}$ be applied at its top surface and let both ends be *simply supported* (i.e., have $u\_y = 0$ at the neutral axis) by a roller support. In addition, both ends are subjected to equal moments that each displaces the beam center downwards (Figure 6.1). The end moment has a value of $M = 1.25e3\,\mathrm{N\text{-}m}$. The material is aluminum 1060. This is a problem where the stresses depend only on the geometry. However, the deflections always depend on the material type.

It should be clear that this problem is symmetrical about the vertical centerline (why that is true will be explained shortly should it not be clear). Therefore, no more than half the beam needs to be considered (and half the load). Select the right half. The beam theory results should suggest that an even more simplified model would be valid due to anti-symmetry (if we assume half the line load acts on both the top and bottom faces). The 3D flat face symmetry restraint was described earlier. The 2D nature of this example provides insight into how to identify lines (or planes in 3D) of symmetry and anti-symmetry, as shown in Figure 6.2.

A process for identifying displacement restraints on planes of symmetry and anti-symmetry will be outlined here. Assume that the horizontal center line of the beam corresponds to the dashed centerline of the anti-symmetric image at the left in Figure 6.3.

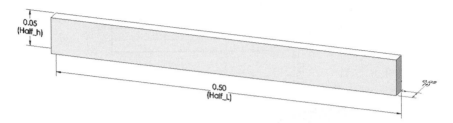

Fig. 6.2.   One-quarter of the beam.

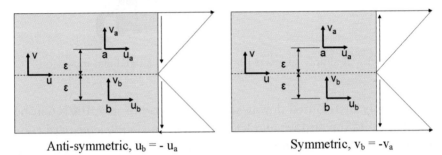

Anti-symmetric, $u_b = -u_a$ Symmetric, $v_b = -v_a$

Fig. 6.3.   Anti-symmetric ($u = 0$, $v = ?$), and symmetric ($u = ?$, $v = 0$) displacement states.

The question is, what, if any, restraint should be applied to the u or v displacement component on that line.

To resolve that question imagine two mirror image points, a and b, each a distance, ε, above and below the dashed line. Note that both the upper and lower half portions are loaded downward in an identical fashion, and they have the same horizontal end supports. Therefore, you expect v_a and v_b to be equal, but have an unknown value (say $v_a = v_b = ?$). Likewise, the horizontal load application is equal in magnitude, but of opposite sign in the upper and lower regions. Therefore, you expect $u_b = -u_a$. Now let the distance between the points go to zero ($\varepsilon \to 0$). The limit gives $v = v_a = v_b = ?$, so v is unknown and no restraint is applied to it. The limit on the horizontal displacement gives $u = u_b = -u_a \to 0$, so the horizontal displacement can be restrained to zero if you with to use a half depth anti-symmetric model. Another way to say that is: *on a line or plane of anti-symmetry the tangential displacement component(s) is restrained to zero*. For a shell or beam the rotational component normal to an anti-symmetry plane is also zero.

The vertical centerline symmetry can be justified in a similar way. Imagine that the right image in Figure 6.3 is rotated 90 degrees clockwise so the dashed line is parallel to the beam vertical symmetry line. Now u represents the displacement component tangent to the beam centerline (i.e., vertical). The vertical loading on both sides is the same, as are the vertical end supports, so the vertical motion at a and b will be the same (say $u_a = u_b = ?$). In the limit, as the two points approach each other $u = u_a = u_b = ?$, so the beam vertical centerline has an unknown tangential displacement and is not subject to a restraint. Now consider the displacement normal to the beam vertical centerline (here v). At any specified depth, the loadings and deflections in that direction are equal and opposite. Therefore, in the limit as the two points approach each other $u = u_b = -u_a \rightarrow 0$, so the displacement component normal to the beam vertical centerline must vanish. Another way to state that is: *on a line or plane of symmetry the normal displacement component is restrained to zero*. For a shell or beam the rotational components parallel to a symmetry plane are also zero.

From the above arguments, the 2D approximation can be reduced to one-quarter of the original domain. The other material is removed and replaced by the restraints that they impose on the portion that remains. Now your attention can focus on the applied load states. The top (and bottom) line load can be replaced either with a total force on the top surface, or an equivalent pressure on the top surface, since SW Simulation does not offer a load-per-unit-length option. Unfortunately, either requires a hand calculation that might introduce an error. The less obvious question is how to apply the end moment(s).

Since the general shell element has been forced to lie in a flat plane, and have no loads normal to the plane, its two in-plane rotational DOF will be identically zero. However, the nodal rotations normal to the plane are still active (in the literature they are call drilling freedoms in 2D studies). That may make you think that you could apply a moment, M_z, at a node on the neutral axis of each end of the beam. In theory, that should be possible, but in practice it works poorly (try it) and the end moment should be applied in a different fashion. One easy way to apply a moment is to form a couple by applying equal positive and negative triangular pressures across the depth of the ends of the beam. That approach works equally well for 3D solids that do not have rotational degrees of freedom.

The maximum required pressure is related to the desired moment by simple static equilibrium. The resultant horizontal force for a linear pressure variation from zero to p_{max} is $F = A\,p_{max}/2$, where A is the corresponding rectangular area, $A = t(h/2)$, so $F = thp_{max}/4$. That resultant force occurs at the centroid of the pressure loading, so its lever arm with respect to the neutral axis is $d = 2(h/2)/3 = h/3$ (for the top and bottom portions). The pair of equal and opposite forces form a combined couple of $M_z = F(2d) = th^2p_{max}/6$. Finally, the required maximum pressure is $p_{max} = 6M_z/th^2$.

To apply this pressure distribution in SW Simulation you must define a local coordinate system located at the neutral axis of the beam and use it to define a variable pressure. However, the SW Simulation non-uniform pressure data requires a pressure scale, p_{scale}, times a non-dimensional function of a selected local coordinate system. Here you will assume a pressure load linearly varying with local y placed at the neutral axis: $p(y) = p_{scale} * y$ (with y non-dimensional). This must match p_{max} at $y = h/2$, so

$$p_{scale} = \frac{2p_{max}}{h} = \frac{12M_z}{th^3}.$$

It is often necessary to apply moments to solids in this fashion. This moment loading will be checked against beam theory estimates before applying the line load. Here, $p_{max} = 3.75e7\,\text{N/m}^2$, $p_{scale} = 7.5e8$.

The beam theory solution for a simply supported beam with a uniform load is well known, as is the solution for the loading by two end moments (called pure bending). In both cases the maximum deflection occurs at the beam mid-span. The two values are $v_{max} = 5wL^4/384EI$, and $v_{max} = ML^2/8EI$, respectively. Here the centerline deflection due only to the end moment is $v_{max} = 1.36e\text{-}3\,\text{m}$. For a linear analysis and the sum of these two values can be used to validate the centerline deflection. Next, the one-quarter model, shown in Figure 6.2, will be built, restrained, and loaded:

(1) Build the rectangle sketch and convert it to a planar surface via **Insert → Surface → Planar Surface** and select the Current Sketch as the Planar Surface.

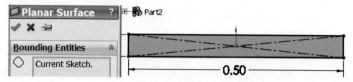

(2) Start a new study using a shell mesh: **Simulation → New Study → Static**. Name it Anti-symm-beam.
(3) Use **Part → Edit Definition** to set the element **Type** to Thin, and the **Shell Thickness** to be 0.02 m. Also use **Part → Apply/Edit Material** to select the library material of 1060 aluminum.

Since the stresses through the depth of the beam are going to be examined here, you should plan ahead and insert some split lines on the front surface to be used to list and/or graph selected stress and deflection components:

(1) Right click on the front face, **Insert Sketch**.
(2) Insert a **line** segment that crosses the face at the interior quarter points. Including the end lines, five graphing sections will be available. Also add a right corner arc for the vertical support edge.

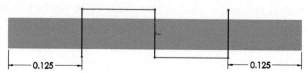

(3) **Insert** → **Curve** → **Split Line** select the body faces, click **OK**.

Remember that shells defined by planar surfaces must have their restraints and loads applied directly to the edges of the selected surface. First the symmetry and anti-symmetry restraints will be applied. Since the shell mesh will be flat it is easy to use its edges to define directions for loads, or restraints:

(1) Right clicking on **Fixtures** opens the **Fixture panel**.
(2) The zero horizontal (x) deflection is applied as a vertical symmetry condition on the edge corresponding to the beam centerline; **Advanced** → **Use Reference Geometry**, select vertical **Edge 1** to restrain and horizontal **Edge 2** for the direction.

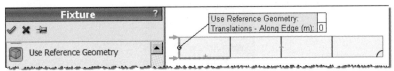

(1) Apply the anti-symmetry condition along the edge of the neutral axis. **Use reference geometry**, select the five bottom edges formed by the split lines and **Edge 7** for the direction.

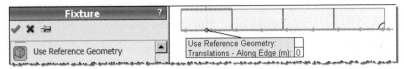

At the simply supported end it is necessary to assume how that support will be accomplished. Beam theory treats it as a point support, but in 2D or 3D that causes a false infinite stress at the point. Another split line arc was introduced so about one-third of that end could be picked to provide the vertical restraint required. This serves as a reminder that where, and how, parts are restrained is an assumption. So it is wise to investigate more than one such assumption. Software tutorials are intended to illustrate specific features of the software, and usually do not have the space for, or intention of, presenting the best engineering judgment. Immovable restraints are often used in tutorials, but they are unusual in real applications. Apply the right vertical end support restraint:

(1) Select **Fixtures** to open the **Fixture panel**. Pick **Advanced** → **Use Reference Geometry**.
(2) Select the lower right front vertical edge line to restrain, and the upper vertical edge as the direction.

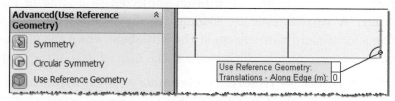

Since a general shell element is being used in a plane stress (membrane shell) application it still has the ability to translate normal to its plane and to rotate about the in-plane axes (x and y). Those three rigid body motions should also be eliminated in any plane stress analysis on most FEA systems. If nothing else, such a restraint avoids calculating three zero values at each node and makes your analysis more efficient. At worst, during the solve phase you may get a fatal error message (due to round-off errors):

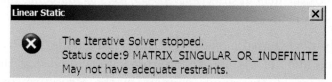

For good modeling practice apply those restraints via:

(1) Select **Fixtures** to open the **Fixture panel** and pick **Advanced Fixtures**.
(2) Pick **Symmetry**; select the five front face segments of the beam to restrain. Change the **Symbol Color**. That completes the symmetry, anti-symmetry and rigid body motion restraints for this model.

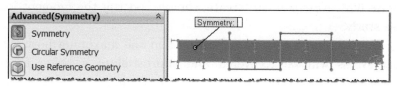

Unlike general shells, membrane shells do not have active rotational degrees of freedom that allow for the direct application of a couple. A linear variation of equal and opposite pressures, relative to the neutral axis, can be used to apply a statically equivalent moment to a continuum body that does not have rotational degrees of freedom. Such a loading also has the side benefit of matching the theoretical normal stress distribution in a beam subjected to a state of pure bending. A varying pressure loading usually requires the user to define a local coordinate system at the axis about which the moment acts. In this case, it must be located at the neutral axis of the beam:

(1) Select **Insert → Reference Geometry → Coordinate System** to open the **Coordinate System panel**. Right click on the right end of the neutral axis to set the **origin** of Coordinate System 1.

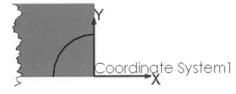

(2) Accept the default directions for the axes as matching the global axes.

To apply the external moment use **External Loads → Pressure**. The application of the non-uniform pressure is applied at the front vertical edge at the simple support in the **Pressure panel** of Figure 6.4. A unit **pressure value** is used to set the units and the magnitude is defined by multiplying that value by a non-dimensional polynomial of the spatial coordinates of a point, relative to local **Coordinate System 1** defined above.

Having completed the restraints and moment loading, the default names in the manager menu have been changed (by slow double clicks) to reflect what they are intended to accomplish (left of Figure 6.5). Now you can create a mesh and run the moment load case study.

The maximum vertical deflection and the maximum horizontal fiber stress will be recovered and compared to a beam theory

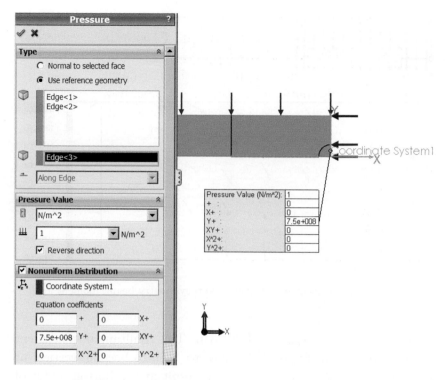

Fig. 6.4. Applying the required linear pressure.

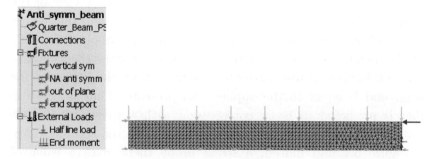

Fig. 6.5. Beam quarter model membrane shell mesh.

estimate in order to try to validate the FEA study. For this simple geometry and pure bending moment the beam theory results should be much more accurate than is usually true. As stated above, the maximum vertical deflection at the centerline is predicted by thin

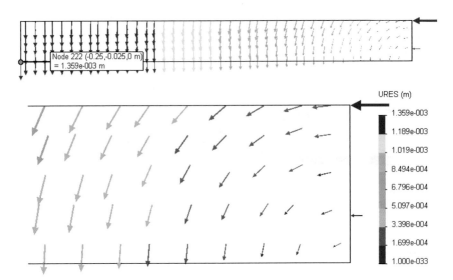

Fig. 6.6. Displacement in the quarter anti-symmetry beam, moment load.

beam theory to be $v_{max} = 1.36e\text{-}3$ m. The resultant displacement vectors are seen in Figure 6.6. They are seen to become vertical at the centerline. A probe displacement result at the bottom point of the vertical centerline line gives $v_{FEA} = 1.36e\text{-}3$ m, which agrees to three significant figures with the elementary theory. A view of the support region (bottom of Figure 6.6) shows that the displacement vectors close to the restraint are basically rotating about that end.

For bending by end couples only, the elementary theory states that the horizontal fiber stress is constant along the length of the beam and is equal to the applied end pressure. That is, the top fiber is predicted to be in compression with a stress value $\sigma_x = p_{max} = 3.75e7$ N/m^2. That seems to agree with the contour range in Figure 6.7 and indeed, a stress probe there gives a value of $\sigma_x = -3.77e7$ N/m^2. Beam theory gives a linear variation, through the depth, from that maximum to zero at the neutral axis.

To compare with that, a graph of along the quarter point split line is given in Figure 6.8. It shows that the seven nodes along the edges of the three quadratic elements have picked up the predicted linear graph quite well. For the next load case of a full span line

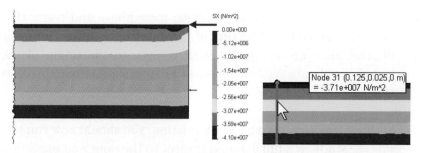

Fig. 6.7. Horizontal stress at right L/8 span segment and its probe value.

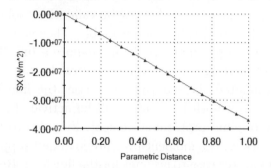

Fig. 6.8. Graph of horizontal stress at vertical line L/8 from the support.

load the shear stress (that is zero here) will be parabolic and the corresponding graph will be less accurate for such a crude mesh.

For this first load case, the only external applied load is the horizontal pressure distribution. It caused a resultant external horizontal force that was shown above to be $F = 18{,}750\,\text{N}$. You should expect the finite element reaction to be equal and opposite of that external resultant load. Check that in the Manager menu:

(1) Right click **Results** → **List Result Force** → **Reaction Force** to open the panel with the reaction forces.

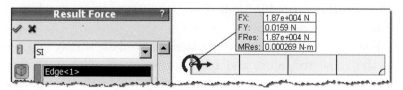

(2) Examine the horizontal (x) reaction force above and verify that its sum is 18,750 N. (The sum of the moments is often confusing because they are computed with respect to the origin of the global coordinate system, and most programs never mention that fact.)

To test your experience with SW Simulation, you should now run this special case study as a full 3D solid subject to the same end pressures. You will find this model was quite accurate. While planning 3D meshes you can get useful insights by running a 2D study like this. Also, a 2D approximation can be a useful validation tool if no analytic results or experimental values are available. They can also be easier to visualize. Of course, many problems require a full 3D study but 1D or 2D studies along the way are educational.

Having validated the moment load case, the line load will be validated and then both load cases will be activated to obtain the results of the original problem statement. First, go to the manager menu, right click on the moment pressure load and suppress it. Next you open a new force case to account for the line load. Recall that the line load totaled 10,000 N. Since the part has been reduced to one-fourth, through the use of symmetry and anti-symmetry, you only need to distribute 2,500 N over this model. There are two ways to do that for selected surface shell formulation of any plane stress problem. They are to apply that total as either a line load, or to distribute it over the mesh face as a tangential shear traction (which is the better way). Figure 6.9 (left) shows the Apply Force approach. That approach has been made less clear by the way the split lines were constructed. The top of the beam has been split into four segments and this method applies a force *per entity*. Therefore, a resultant force of 625 N per edge segment is specified. Had the split lines not had equal spacing you would have to measure each of their lengths and go through this procedure four times (the pressure approach avoids that potential complication).

With this second load case in place the study is simply run again with the same restraints and mesh. A series of quick spot checks of the results are carried out before moving on to the true problem

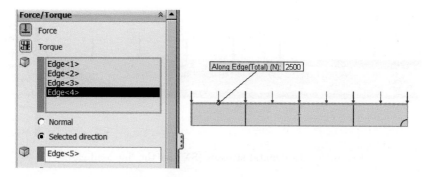

Fig. 6.9.   Second beam load case of a line load.

where both load cases are activated. The beam theory validation result, for this line load, predicted a maximum vertical centerline deflection of $v_{max} = 1.13e\text{-}3$ m. The plane stress maximum deflection was extracted:

(1) Double click on Plot1 under displacements. The contoured magnitude shows a rotational motion about the simple support end, and vertical translation at the beam centerline, as expected.

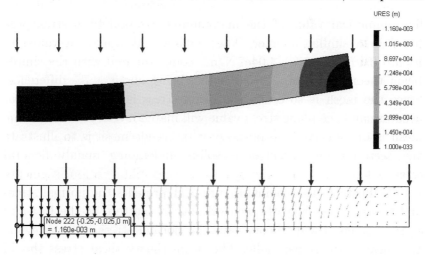

(2) Right click in the manager menu **Results** → **List Stress, Displacement, Strain** to open the **List Results panel**. Select **Displacements** and under **Advanced Options** select **Absolute Max**, and **Sort by value**.

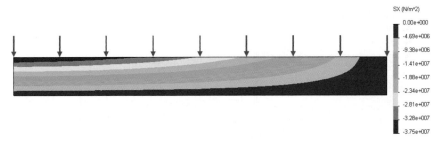

SX (N/m^2)

0.00e+000

-4.69e+006

-9.38e+006

-1.41e+007

-1.88e+007

-2.34e+007

-2.81e+007

-3.28e+007

-3.75e+007

Fig. 6.10. Horizontal stresses (SX) for the line load case.

(3) When the list appears note that the maximum deflection is
1.16e-3 m at the centerline position. That is very close to the
initial validation estimate.

Node	X (m)	Y (m)	Z (m)	URES (m)	▲
222	-0.25	-0.025	0	1.15965e-003	
2109	-0.25	-0.021875	0	1.15963e-003	
2106	-0.246875	-0.025	0	1.15958e-003	
221	-0.25	-0.01875	0	1.15957e-003	
2105	-0.246875	-0.021875	0	1.15956e-003	

The numerical value of the maximum horizontal fiber stress was
listed in a similar manner. The maximum compression value, in
Figure 6.10, of $\sigma_x = -4.04e7 \, \text{N/m}^2$ compares well with the simple
beam theory value of $-3.75e7 \, \text{N/m}^2$, being about a 7% difference.
Since the mesh is so crude the beam stress is probably the most
accurate and the plane stress value will match it as a reasonably fine
mesh is introduced. The purpose of the crude mesh is to illustrate
the need for mesh control is solids undergoing mainly flexural
stresses.

To illustrate that point, Figure 6.11 presents the normal stress
and shear stress, from the neutral axis to the top, at the L/4 and
L/8 positions. Beam theory says the normal stress is linear while
the shear stress is parabolic. The beam theory shear stress should
be zero at the top fiber and, for a rectangular cross-section, has
a maximum value at the neutral axis of $1.88e6 \, \text{N/m}^2$. The graph
values in Figure 6.11 shows a plane stress maximum shear stress
of $1.84e6 \, \text{N/m}^2$ and a minimum of $0.08 \, e6 \, \text{N/m}^2$ at the quarter span

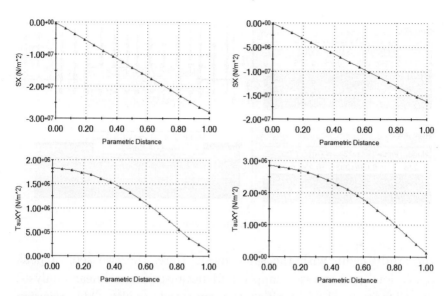

Fig. 6.11. Normal and shear stresses at L/4 (left) and L/8 for the line load.

section. That is quite good agreement with a validation estimate from beam theory.

6.3. Combining Load Cases

Having validated each of the two load cases they are combined by un-suppressing the end moment condition (Figure 6.12 left) and running

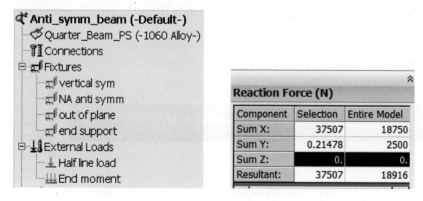

Fig. 6.12. Verifying the model reactions for the combined loadings.

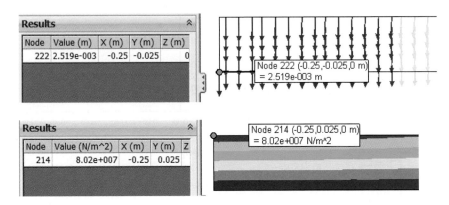

Results   ≫

Node	Value (m)	X (m)	Y (m)	Z (m)
222	2.519e-003	-0.25	-0.025	0

Node 222 (-0.25,-0.025,0 m)
= 2.519e-003 m

Results   ≫

Node	Value (N/m^2)	X (m)	Y (m)	Z
214	8.02e+007	-0.25	0.025	

Node 214 (-0.25,0.025,0 m)
= 8.02e+007 N/m^2

Fig. 6.13.   Combined loading displacement and von Mises stress.

the study again with the same mesh. Here, the two sets of peak deflections and stresses simply add because it is a linear analysis. A quick spot check verifies the expected results. The reaction force components were verified (Figure 6.12 right) before listing the maximum deflection and fiber stress (Figure 6.13).

What remains to be done is to examine the likely failure criteria that could be applied to this material. They include the von Mises effective stress, the maximum principle shear stress, and the maximum principle normal stress. The von Mises contour values are shown in Figure 6.14. Twice the maximum shear stress (the stress intensity) is given in the top of Figure 6.15, while the bottom portion displays the maximum principle stress. Actually, the principal stress, P3, is compressive here but it corresponds to the mirror image tension on the bottom fiber of the actual beam.

von Mises (N/m^2)

- 8.02e+007
- 6.70e+007
- 5.38e+007
- 4.05e+007
- 2.73e+007
- 1.40e+007
- 7.88e+005

Fig. 6.14.   von Mises stress in the beam with a line load.

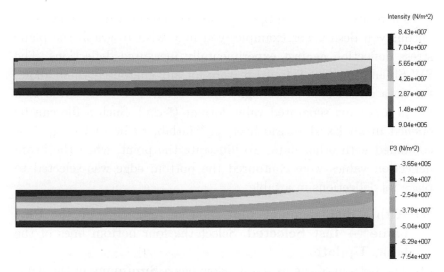

Fig. 6.15.   Beam principal stress and maximum shear contours.

All three stress values need to be compared to the yield point stress of $2.8e7\,\text{N/m}^2$. The arrow in the figure highlights where that falls on the color bar. All of the criteria exceed that value, so the part will have to be revised. At this point failure is determined even before a material Factor of Safety has been assigned. For ductile materials, the common values for the material FOS range from 1.3 to 5, or more [9, 12]. Assume a FOS = 3. The current design is a factor of about 3.3 over the yield stress. Combining that with the FOS means that the stresses need to be reduced by about a factor of 10.

The cross-sectional moment of inertia, $I = th^3/12$, is proportional to the thickness, t, so doubling the thickness cuts the deflections and stresses in half. Changing the depth, h, is more effective for bending loads. It reduces the deflection by $1/h^3$ and the stresses by a factor of $1/(2\,h^2)$. The desired reduction of stresses could be obtained by increasing the depth by a factor of 2.25. The above discussion assumed that buckling has been eliminated by a buckling analysis. Since buckling is usually sudden and catastrophic it would require a much higher FOS.

There are times when the software will not provide the graphical output you desire. For example, you may wish to graph the plane stress deflection against experimentally measured deflections. The SW Simulation *List Selected* feature for any contoured value allows the data on selected edges, split lines, or surfaces to be saved to a file in a comma separated value format (*.csv). Such a file can be opened in an Excel spreadsheet, or Matlab, to be plotted and/or combined with other data. To illustrate the point, when the beam deflection values were contoured the bottom edge was selected to place its deflections in a table:

(1) With a **displacement plot** showing right click on the **Plot name** → **List Selected**. Select the four bottom lines of the beam. **Update**.

(2) The bottom of the listing window has a **Summary** of the data.

Summary		
	Value	
Sum	0.31713	m
Avg	0.0019697	m
Max	0.002529	m
Min	1.1363e-007	m
RMS	0.0020401	m

(3) The lower **Report Option** region does not include the Graph Icon, but does show a **Save icon**. That is because the path has multiple lines. Pick **Save** to have the listed data (node number, deflection value, and x-, y-, z-coordinates) to be output as a comma separated values (csv) file.

(4) **Name and save** the data for use elsewhere.

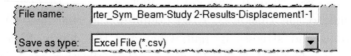

File name:	rter_Sym_Beam-Study 2-Results-Displacement1-1
Save as type:	Excel File (*.csv) ▼

SW Simulation did not offer a plot option along all the selected lines since could not identify which item to sort. You know that the multiple line segments should be sorted by the x-coordinate value. Therefore, the data were opened in Excel, sorted by x-coordinate, and graphed as deflection versus position (Figure 6.16). You could add experimental deflection values to the same file and add a second

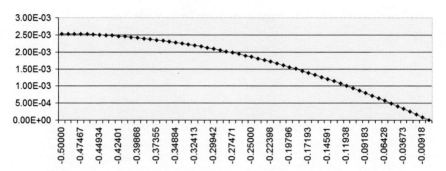

Fig. 6.16.   Excel graph from saved CSV file for combined load.

curve to the display for comparison purposes. Other aspects of plane stress analysis are covered in the next section on rotational loadings, and later sections on 2.5D studies, and buckling.

7

Rotational Loads and Accelerations

7.1. Introduction

This example will look at essentially planar objects subjected to centrifugal loads. That is, loads due to angular velocity and/or angular acceleration about an axis. The part under consideration is a spinning grid strainer that rotates about a center axis perpendicular to its plane. The part has five symmetrical segments, of 72 degrees each, and each segment has a set of slots that have mirror symmetry about a plane at 36 degrees. The questions are: (1) Does a cyclic symmetric part, with respect to its spin axis, have a corresponding set of cyclic displacements and stresses when subject to an angular velocity, ω, about that axis? (2) Does it have the same type of behavior when subjected to an angular acceleration?

To answer these questions you need to recall the acceleration kinematics of a point mass, $dm = \rho dV$, following a circular path of radius r. In the radial direction there are usually two terms, $r\omega^2$ that always acts toward the center and d^2r/dt^2 acting in the direction of change of the radial velocity. The latter term is zero when r is constant, as on a rigid body. In the tangential direction there are also two components in general: $r\alpha$ acting in the direction of α and a Coriolis term of $2dr/dt\omega$, in the direction of ω if dr/dt is positive. Again the latter term is zero for a rigid body. The remaining radial acceleration $(r\omega^2)$ always acts through the axis of rotation (as a purely radial load), thus, it will have full cyclic symmetry. Angular acceleration always acts in the tangential direction a rotating part, as it spins up or spins down. The worst case is often at a sudden stop where a large angular velocity exists and a negative large angular acceleration is applied.

A part with angular acceleration is basically subjected to torsional or cantilever like loading about its axis of rotation. To illustrate these concepts consider a cantilever beam rotating about an axis outside its left end. The deformed shapes and stress levels for constant angular velocity were seen in Figures 3.19 through 3.21. The angular acceleration results for a similar cantilever will be considered next. For a radial spoke, the angular acceleration effect is similar to a transverse linearly increasing line load acting on a cantilever beam. The transverse force would equal the mass per unit length of the spoke times the tangential acceleration, $f = \rho A r \alpha$. Simple beam theory would thus predict an end deflection of $v = 11 \, m \alpha L^4 / 120 \, EI$ for a spoke mass of $m = \rho A L$. The simple spoke model in Figures 7.1 and 7.2 agrees very closely with that deflection estimate. The von Mises stress (Figure 7.2) shows a typical bending response about the radial centerline. An extension of this model will be used to estimate the expected angular acceleration loading on the more geometric complex part considered in the next section.

7.2. Building a Segment Geometry

The desired rotating spin_grid is shown in Figure 7.3. The part has a inner shaft hole diameter, and outermost diameter of 1 inch and

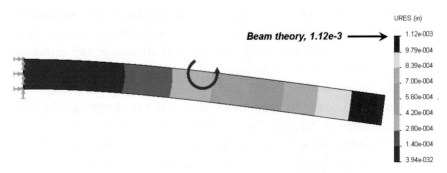

Fig. 7.1.   Radial spoke under angular acceleration.

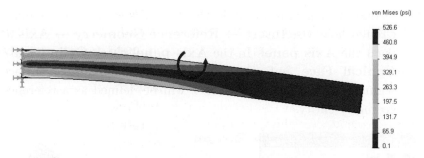

Fig. 7.2.   von Mises stress in a spoke with angular acceleration about left end.

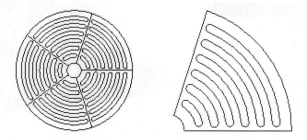

Fig. 7.3.   Full geometry and its one-fifth symmetry.

4 inches, respectively. It has six curved slots 1/4 inch wide, symmetric about the 36 degree line, and ends in a arc that is 1/8 inch from the 0 degree line and is 0.20 inches thick.

Prepare a sketch in the top view with several radial and arc construction lines:

(1) **Front → Insert Sketch**, build several construction lines for the center of the slots, the symmetry plane, and an off-set horizontal line for the fillet centers. Add arcs, and line segments to close the shape. **Extrude**, about the mid-plane, to the specified thickness.

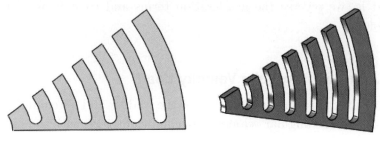

(2) Define the axis of rotation. In this case it is the axis of the inner circular hole. Use **Insert → Reference Geometry → Axis** to open the **Axis panel**. In the **Axis panel** check **Cylindrical/ Conical Face** and select the inner-most cylindrical surface segment of the part and **Axis 1** will be defined as a reference geometry entity.

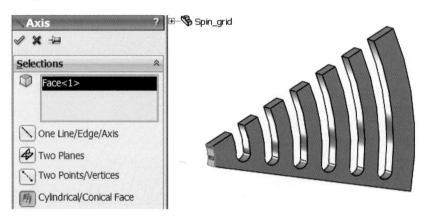

At this point the smallest geometrical region is complete and you can move on the **SW Simulation Feature Manager** to conduct the deflection and stress analysis.

To open a centrifugal analysis you have to decide if it is classified as static, vibration, or something else. Instead of applying Newton's second law, $\boldsymbol{F} = m\boldsymbol{a}$, for a dynamic formulation (where is $\boldsymbol{F}$ the resultant external force vector and is $\boldsymbol{a}$ the acceleration vector of mass m), D'Alembert's principle is invoked to use a static formulation of $\boldsymbol{F} - \boldsymbol{F}_I = \boldsymbol{0}$, where the inertia force magnitude is $F_I = mr\omega^2$ in the radial direction and/or $mr\alpha$ in the tangential direction. That is, we reverse the acceleration terms and treat it as a static problem.

7.3. Part with Angular Velocity Load

Since this part has symmetry in the radial and circumferential directions, the angular velocity case can be obtained with the smaller

36 degree segment. In SW Simulation:

(1) Right click on the **Part_name** → **Study, Name** the part (spin_grid here), pick **Static** analysis. The study defaults to solid elements. (A planar model is almost always the cheapest and fastest way to do initial studies of a constant thickness part.)

(2) When the **Study Menu** appears right click on **Solid** → **Edit/Define Material**.

(3) In the **Material panel** pick **Library** → **Steel** → **Cast Alloy**. Select **Units** → **English**. Note that the yield stress, is about $35e3$ psi. It will be compared to the von Mises Stress failure criterion later. (The mass density, is actually weight density, $\gamma = \rho g$, and is mislabeled.)

For the centrifugal load due to the angular velocity only the spoke center plane and slots center plane always move in a radial direction. That means that they have no tangential displacement (that is, no displacement normal to those two flat radial planes). In the Manager menu:

(1) Select **Fixtures** to activate the first **Fixture panel**. Pick **Advanced** → **Symmetry** as the **Type** and select the part face in the 0 degree plane. That sets the normal displacement component to zero. Preview the restraints, and then click **OK**.

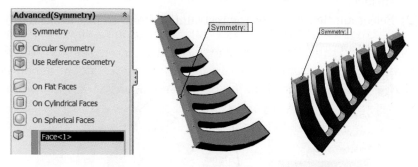

(2) Repeat that process and select all seven of the part flat faces lying in the 36 degree plane (above).

At this point a total of eight surfaces are required to have only radial displacements. Either of the above two restraint operations eliminates a possible rigid body rotation about the axis of rotation. The inner arc of the grid is assumed to be force fit bonded to the shaft in the circumferential direction. Use **Fixtures → Advanced → On Cylindrical Faces → Circumferential**.

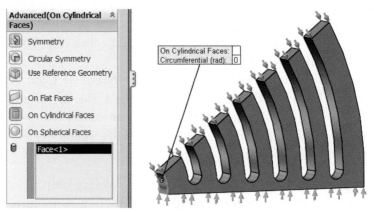

To prevent a rigid body translation of the part in the direction parallel to the axis you should also at least restrain the cylindrical shaft contact surface in that direction as well. An alternative, for this loading, is to use symmetry of the front or back face of the part.

(1) Select **Fixtures** to activate the first **Fixture panel**. Pick **Advanced → Symmetry**, select the inter-most cylindrical surface.

(2) Select another color for the restraints present to just avoid RBM.

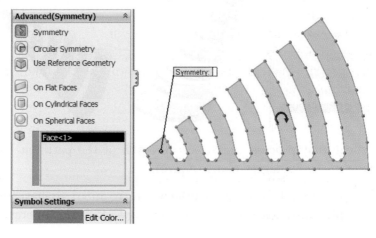

Next you set the centrifugal body force loads due to the angular velocity. That will require picking the rotational axis so select **View → Axes** first and then:

(1) Right click **External Loads → Centrifugal** for the **Centrifugal panel**.
(2) There pick **Axis 1**, set rpm as the **Units** and type in 1,000 rpm for the **angular velocity**.

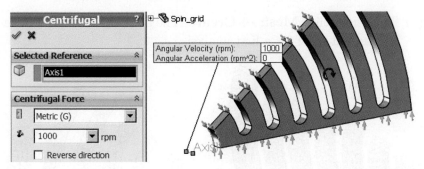

The radial acceleration, $a_r = r\omega^2$, varies linearly with distance from the axis, so expect the biggest loads to act on the outer rim. Also remember that the radial acceleration is also proportional to the square of the angular velocity. Thus, after this analysis if you want to reduce the stresses by a factor 4 you cut the angular velocity in half. (You would not have to repeat the analysis; just note the scaling in your written discussion. But you can re-run the study to make a pretty picture for the boss.)

For this preliminary study each curved ring segment will act similar to straight fixed-fixed beam of the same length under a transverse gravity load. (To know about what your answers should be from SW Simulation, do that simple beam theory hand calculation to estimate the relative deflection at the center as well as the center and end section stresses.) Thus, bending stresses may concentrate near the ends so make the mesh smaller there:

(1) Right click **Mesh → Apply Control** to activate the **Mesh Control panel**. There pick the six bottom arc faces as the **Selected Entities** and specify an element size.

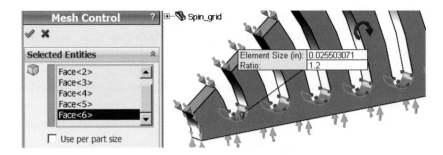

(2) Then right click **Mesh → Create**. The initial mesh looks a little coarse, but okay for a first analysis.

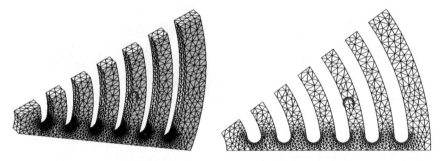

(3) Start the equation solver by right clicking on the study name and selecting **Run**. When completed review the results.

Both the displacements and stresses should always be checked for reasonableness. Sometimes the stresses depend only on the shape of the material. The deflections always depend on the material properties. Some tight tolerance mechanical designs (or building codes) place limits on the deflection values. SW Simulation deflection values can be exported back to SolidWorks, along with the mesh model, so that an interference study can be done in SolidWorks. You should always see if the displacements look reasonable:

(1) Double click on **Results → Define Displacement Plot** to see their default display, which is a continuous color contour format. Such a pretty picture is often handy to have in a report to the boss, but the author believes that more useful information is conveyed with the discrete band contours.

(2) Right click on **Plot 1** → **Settings** → **Fringe Options** → **Discrete** for the default deformed shape contours.

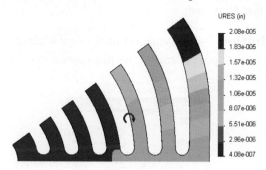

(3) Since displacements are vector quantities you convey the most accurate information with vector plots. Right click **Plot 1**; **Edit Definition** → **Advanced** → **Show as a vector plot**.

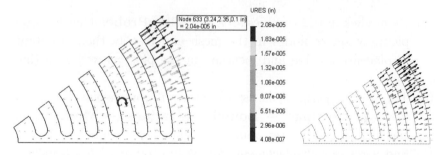

(4) When the vector plot appears, dynamically control it with **Vector Plot Options** and increase **Size**. Then dynamically vary the **Density** (of nodes displayed) to see different various nodes and their vectors displayed. Retain the one or two plots that are most informative.

As expected, the center of the outer-most rim has the largest displacement while the smallest displacement occurs along the radial "spoke" centered on the 0 degree plane. You may want to compare the relative displacement of the outer ring to a handbook approximation in order to validate the computed displacement results (and your knowledge of SW Simulation). Localized information about selected displacements at nodes or lines is also available.

For problems with angular velocity or angular accelerations you should also view the displacements in cylindrical coordinate components. For that use:

(1) Edit **Definition** → **Advanced** and select **axis-1** from the part tree. Then the part's radial displacement component will be displayed as UX.

(2) Right click in the graphics area and select **Probe**. That lets you pick any set of nodes in the mesh and display their resultant displacement value and location on the plot, as well as listing them (above).

(3) The probe operation is usually easier if you display the mesh first with **Settings** → **Boundary Options** → **Mesh**. Using the support and center points in Figure 7.4 as probe points you find a relative displacement difference of about 1.65e-5 inches. That can be compared to a fixed-fixed uniformly loaded beam mid-span deflection estimate.

All of the physical stress components are available for display, as well as various stress failure criteria. The proper choice of a failure criterion is material dependent, and it is the user's responsibility to know which one is most valid for a particular material. The so-called Effective Stress (actually distortional energy), or von Mises stress, value is often used for ductile materials.

The von Mises failure criterion (a scalar quantity) is superimposed on the deformed shape in Figure 7.5 using discrete color contours. Also **Chart Options** → **Color Options** was used to select only eight colors.

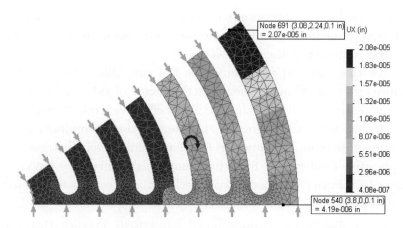

Fig. 7.4. Outer ring radial displacements (UX) at the symmetry planes.

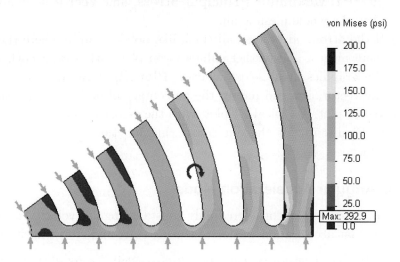

Fig. 7.5. von Mises failure criterion due to angular velocity.

The maximum effective stress is only about 290 psi, compared to a yield stress of about 35,000 psi. That is a ratio of about 134. Since the centrifugal load varies with the square of the angular velocity you would have to increase the current ω by the square root of that ratio (about 11.5). In other words you should expect yielding to occur at $\omega = 11.5$ (1,000 rpm) = 11,500 rpm. Since this is a linear analysis problem, it would not be necessary to repeat the run with that new angular velocity. You could simply scale both the displacements and the stresses by the appropriate constant. If you have a fast computer you may want to do so to include your most accurate plots in your written summary of the analysis.

For this material you would use the von Mises effective stress failure criterion. That is, your material factor of safety is defined as the yield stress divided by the maximum effective stress. It was noted above that the ratio is greater than ten in this preliminary study. The magnitude and direction of the maximum principal stress P1 is informative (and critical for brittle materials). Since they are vector quantities they give a good visual check of the directions of the stress flow, especially in planar studies (they can be quite messy in 3D):

(1) Right click in the graphics area, select **Edit Description** then pick **P1 Maximum Principal Stress**, and **Vector** style and view the whole mesh again.

(2) If the arrows are too small (look like dots) zoom in where they seem biggest and further enhance you plot with a right click in the graphics area, select **Vector Plot Options** and increase the vector size, and reduce the percentage of nodes used for the vector plot. A typical P1 plot, with the deformed shape, is given in Figure 7.6 for the outermost ring junction with the radial spoke.

7.4. Angular Acceleration Model

The previous model considered only constant angular velocity, ω, so the angular acceleration was zero. During start and stop transitions

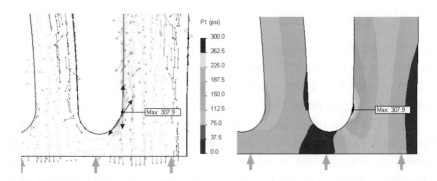

Fig. 7.6. Principal stress P1 at outer ring, due to angular velocity.

both will present and the two effects can be superimposed because this is a linear analysis. Next consider the initial angular acceleration (where $\omega = 0$ for an instant). You can always use the full model, but that takes a lot of computer resources. The most efficient symmetric analysis would require using any 72 degree segment and invoking a special restraint know as circular symmetry, or multiple point constraints or repeated freedoms for nodes on those to edges. That means we know the two edges have the same displacement components normal and tangential to the edges, but they are still unknowns.

You need to identify the axis. Turn on the axes with **View →Axes**. Then select **External Loads → Centrifugal** and pick **Axis 1** in the **Centrifugal panel** and type in the value of the angular acceleration.

7.5. Proper Cyclic Symmetry Model

The above studies show that you can often pick symmetry regions in rotational loadings and drastically reduce the computer resources required. Here it seems like you could refine the outer spoke mesh further to see worst angular velocity effects and refine the innermost gap ends to see the worst angular acceleration effects. Actually, the full model can be replaced by the 72 degree segment using the SW Simulation cyclic (circular) symmetry boundary condition. The automation of a cyclic symmetry analysis requires that the software can express the degrees of freedom in terms of changing coordinate systems established tangential and normal to the repeated surface of cyclic symmetry. That is illustrated in Figure 7.7 (left) where the top view of a cyclic symmetry impeller solid shows some of the pairs of tangential and normal displacements that have to be established and coupled by the analysis software. SW Simulation includes this ability to impose that the normal and tangential displacements on the two limiting surfaces are the same, even though initially unknown. That figure shows that the limiting surfaces do not have to be flat.

An impeller solid body with cyclic symmetry appears in Figure 7.8. Out of plane body parts there would be more influenced by angular acceleration. The two limiting planes do not have to

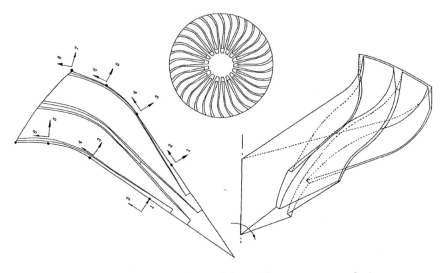

Fig. 7.7.   Impeller well-suited for cyclic symmetry analysis.

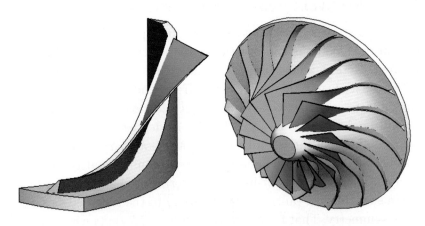

Fig. 7.8.   An impeller solid with 16 segments of cyclic symmetry.

intersect at the geometric center of the body. They only have
to match with themselves when a copy of the included segment
undergoes a rigid body rotation about the geometric center. The
limiting surfaces can occur at changes in the material or changes
in the thickness, like Figure 7.9. There rectangular cyclic symmetry
boundary A–B rotates 60 degrees to join its mating limiting surface

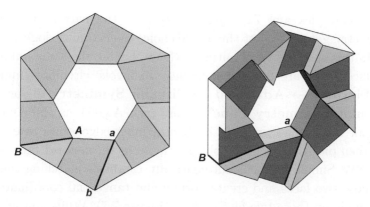

Fig. 7.9. Cyclic symmetry surfaces mate by rotating like rigid bodies.

a–b. Cyclic symmetry conditions also occur in thermal analysis and computational fluid dynamics.

Continuing from the 36 degree segment; use **Insert → Pattern/ Mirror → Mirror** to open the **Mirror panel**. Select the previous part as the **Features to Mirror** and pick one of the free ring faces as the **Mirror Plane, OK**.

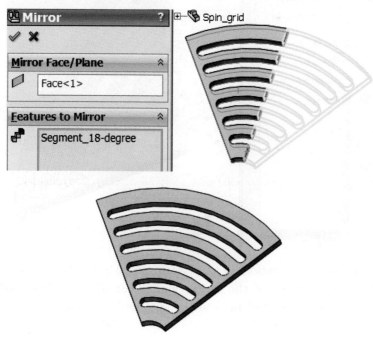

You could pick any 72 degree segment of the part as the cyclic symmetry geometry, but the above choice seemed most logical.

In this case the two limiting surfaces, having the same unknown displacements, are flat radial planes. To invoke circular symmetry use **Fixtures → Advanced → Circular Symmetry** to open the **Circular Symmetry Panel**. There pick **Axis-1** as the common axis for the cyclic symmetry innermost inner edges, pick **one face**, and then pick that face's revolved location as the **second face, OK**. The SW Simulation will automatically establish matching meshes on those two faces and create normal and tangential coordinates at each node on those two faces (like in Figure 7.7). While you see two different nodes at corresponding points on those surface meshes, you should think of them as a single node having normal, and tangential displacement components to be determined by the specific loading conditions.

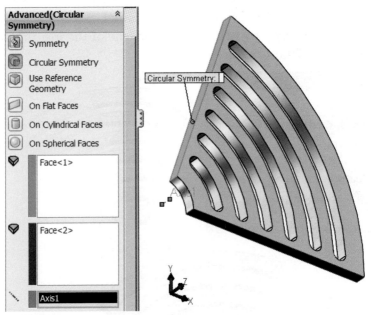

The angular acceleration in this example is input by its fixation to the outer surface of the center shaft (not shown). That would usually be accomplished by a shrink fit, which would also prevent rigid body motion in the axial direction of the shaft. Apply those with **Fixtures → Advanced → On Cylindrical Faces**.

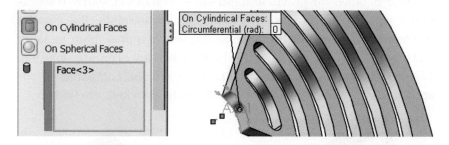

The restraints described above would prevent the six possible rigid body motions. In this case, since the part is flat there will be no axial displacements. Thus, you can reduce the number of unknowns to be solved, and better stabilize an iterative solver, by alternately preventing the z-axis RBM by treating the top face as a symmetry plane. Change the symbol colors to remind yourself that it is more an efficiency restraint as well as restraint against RBM. Use **Fixtures → Advanced → Symmetry**.

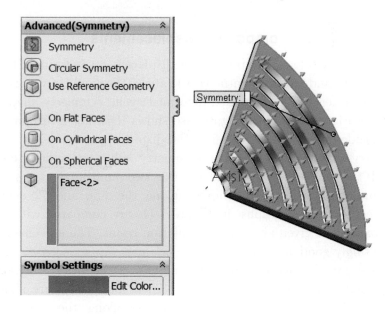

Here you generate solid elements in the full 72 degree segment. The fast iterative solver has worked for crude cyclic symmetry meshes; however it failed for practical mesh sizes. The direct sparse solver is better suited to problems with multipoint constraints, like circular symmetry. It takes much longer to run and requires more memory and disk space but is most likely to yield an accurate solution. The results were obtained with it. They will be compared to the approximate validation predictions from the previous discussions.

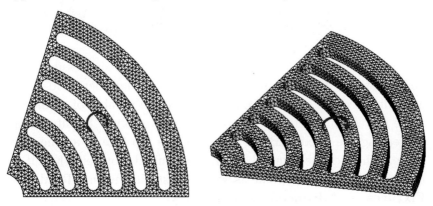

7.6. Cylindrical Component Displacements

Once again, it is logical to present the displacements in terms of their radial (UX) and circumferential (UY) values (the axial UZ values are essentially zero). The amplified deformed cyclic symmetry shape is given in Figure 7.10. That figure also shows the color levels for the magnitude of the main circumferential displacement values.

To verify that the circular symmetry solve has indeed computed the same displacements on the two limiting surfaces the circumferential values are graphed along the zero and 72 degree line segments. The graphs, in Figure 7.11, are compared with the approximation given above as a validation result. The agreement is surprisingly good.

The radial displacements, for angular acceleration only, are about a factor of ten times smaller. Their amplitudes are shown in Figure 7.12. The radial displacements along the outer arc

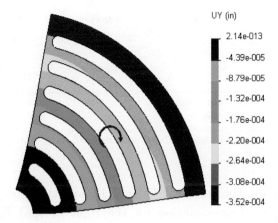

Fig. 7.10.   Amplified deformed shape, with circumferential displacement.

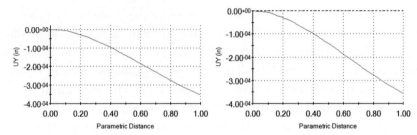

Fig. 7.11. Cyclic symmetry UY values at 72 degrees (top) compared with approximate validation.

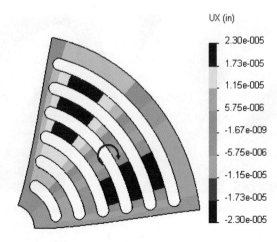

Fig. 7.12.   Radial displacement, UX, is about ten times smaller than UY.

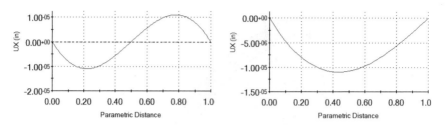

Fig. 7.13. Outer arc radial deflection, compared to validation estimate.

are graphed on the left in Figure 7.13 and are compared with the half model approximation on the right of that figure. Here, the radial displacements are anti-symmetric about the middle 36 degree line.

The von Mises stress is always positive so it shows symmetric distributions (Figure 7.14) under this angular acceleration loads. The levels are quite low and would not govern when compared to the angular velocity results. The maximum principal normal stress, P1, is also quite small, as seen in Figure 7.14. The maximum shear stresses are given in Figure 7.15 and they are also small. All of these possible material failure criterions are directly proportional to the angular acceleration value. They are also within about 20% of

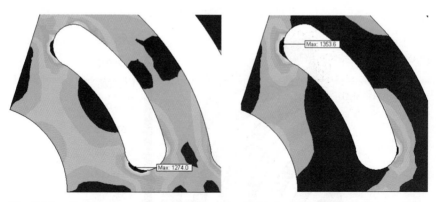

Fig. 7.14. Cyclic symmetry von Mises stress and principal stress P1 from angular acceleration.

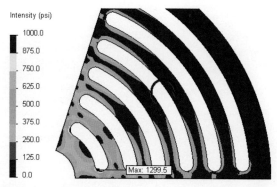

Fig. 7.15.   Twice the maximum shear stress is symmetric for cyclic symmetry restraints.

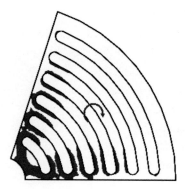

Fig. 7.16.   Main material in the load path for angular acceleration only.

the approximate validation estimates given above. That is unusually good agreement.

The above plots of typical material failure criteria are consistent with the design insight plot in Figure 7.16 of the most active material regions of this part.

8

Flat Plate Analysis

8.1. Introduction

A flat plate is generally considered to be a thin flat component that is subjected to load conditions that cause deflections transverse of the plate. Therefore, the loads are transverse pressures, transverse forces and moment vectors lying in the plane. Those loads are resisted mainly by bending. It is assumed that in-plane membrane stresses are not present and that the transverse displacements are "small". Generally, "small" is taken to mean a deflection that is less than half the thickness of the plate. If the deflection is larger than that and/or membrane forces are present you have to use a non-linear large deflection solution.

8.2. Rectangular Plate

Figure 8.1 shows some of the boundary conditions that can be applied to the edges of a plate. A segment of a plate can be fixed or encastred (left), simply supported (center), or mixed supported (right), or have a free edge. A simply supported condition usually means that the transverse displacement is zero on that segment but the rotation tangent to the segment is unknown. A fixed supported condition usually means that the rotation vector tangent to the segment is also zero. A free edge is stress free. That is, it has no moment or transverse shear resultants acting along its length.

In this section the classic example of a simply supported plate subjected to a uniform transverse pressure will be illustrated.

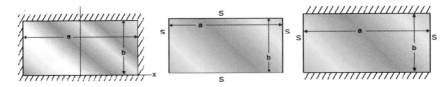

Fig. 8.1.   Some typical boundary condition options on rectangular plates.

Quarter symmetry will be utilized to illustrate symmetry boundary conditions for an element with displacement and rotational degrees of freedom. A short story about this case will be noted at the end. The example plate is AISI 1020 steel with a yield stress of about 51 ksi. The dimensions of the full plate are 4.68 by 12.68 by 0.08 inches and it is subjected to a uniform pressure of 25 psi. The total force is about 371 lb, so you expect the resultant edge reactions to be equal and opposite to that value. Since external edge effects are usually important, a finer mesh is employed along those edges. The plate is set to be of the "thin" type and the study is executed.

The sketch is built and converted to a planar surface with **Insert → Surface → Planar surface**. A static study is opened and the thickness set in **Part Name → Edit Definition → Shell Definition** and it is marked as a thin shell. The two symmetry edges have no in-plane displace normal to the edge, nor any rotation about the edge. They are invoked with **Fixtures → Advanced → Use Reference Geometry**. The two physical support edges are prevented from translation, but can have a rotation vector tangent to the edge. They are set with **Fixtures → Standard → Immovable**. Note that the immovable restraint along a planar curve has the effect of indirectly eliminating the rotation vector normal to the plane as well as the in-plane rotation vector normal to the curve. The two classes of displacement restraints are shown in Figure 8.2.

The constant external pressure is set with **External Loads → Pressure → Normal to**, and the value is set at 25 psi. The loaded model is shown in Figure 8.3. The plate mesh was refined along its edges (which should always be done).

The plate deflections are given in Figure 8.4. The surface deflections are given as contours. The short symmetry edge deflection

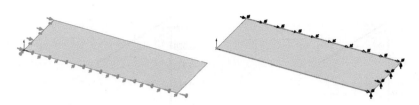

Fig. 8.2.   Symmetry (left) and simple supported plate restraints.

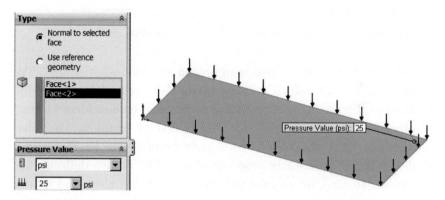

Fig. 8.3.   Constant normal pressure load on the plate.

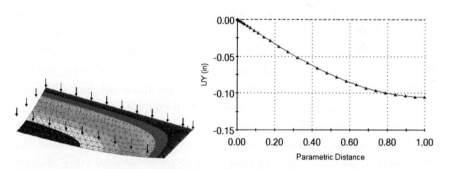

Fig. 8.4.   Deflections of the quarter symmetry plate, and its short edge.

is graphed for more detail in the lower image. The graph starts at the outer (zero deflection) edge and goes to the maximum deflection at the plate center (zero rotation) point. It serves to verify that the restraints were properly applied. The center point deflection can also be compared to analytic estimates [14, 17]. Here the maximum computed deflection is more than half the thickness of the plate,

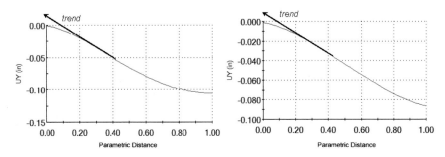

Fig. 8.5.   Deflection along from corner to corner (left) and to mid-side.

therefore the small deflection assumption appears questionable. SW Simulation did not issue a warning about the change in stiffness due to perceived large deflections, but a re-run will be considered later.

Insight to the displacement is obtained by graphing its value along lines from the supported corner point. Note in Figure 8.5 that the trend shows a reverse curvature. The corner deflection is restrained to zero, by an external corner force, but the trend is to a lift up at the corner.

Since plates and shells can be subjected to both bending and membrane (in-plane) stresses the stress results should be checked on the top, bottom, and middle surfaces. Here the membrane stress is zero (*for small deflections*). At a point on the plate the stress will be in tension on one side and have an equal amount of compression on the other. That is important when the material has different strengths in tension and compression (like concrete).

The von Mises effective stress is proportional to the square root of the sum of the squares of the differences in the principal stresses, so it is always positive. The contour, and short symmetry edge, values of the von Mises stress are given in Figure 8.6. Note that the peak values exceed the yield stress, and the material factor of safety with respect to material failure is less than unity.

8.3. Surprising Corner Reactions

The reactions can be recovered in at least two ways. The approach using a free body diagram calculation is shown here. First the

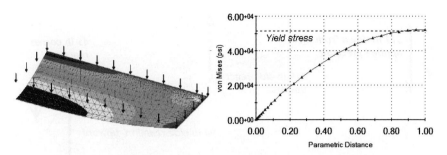

Fig. 8.6.   von Mises stress in the plate and its short symmetry side.

transverse reaction force on the full two supporting edges are recovered and found to be equal and opposite to the resultant applied force from the pressure.

(1) Right click on **Results → List Free Body Force** to open the **Reaction Force panel**.
(2) First select all four supporting edges to get the total reaction forces. The total (above) is about 371 lb, which is equal and opposite to the applied resultant force. Note however, that the corner reaction forces have a negative sign. That is, they act downward in the same direction of the pressure. The total corner reaction force has a value of about 20% of the applied force from the pressure (the corner node force was counted twice).

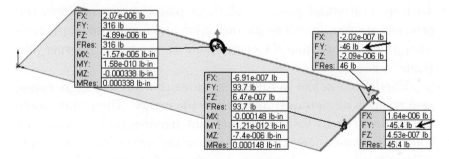

The resultants on a free body diagram come from the integral of the reaction force per unit length of the edge restraints. The reaction per unit length is not constant and will vary in a complicated fashion. Knowing this, the support edges were split to introduce shorter edges

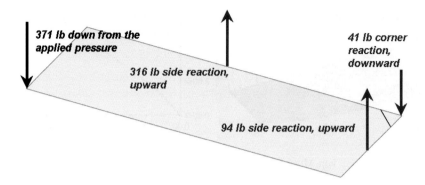

371 lb down from the applied pressure

316 lb side reaction, upward

94 lb side reaction, upward

41 lb corner reaction, downward

Fig. 8.7.   Approximate resultant force and reactions on the plate.

at the corner. That lets you find the portion of the reactions coming from the small corner segments (Figure 8.7).

Now for the related side story: A large analysis group had run the above problem to test a new finite element system that they had recently installed. I was called in as a consultant to fix an "error" they had found. Specifically, when a pressure load was applied downward to a flat plate some of its reactions were also found to be acting downward, just as noted above. That seemed to them to be physically impossible. I stated that the software was giving the proper type of response since elementary plate and shell theory shows that the edge reactions per unit length must behave in that fashion. To help understand why, I had them plot the two non-zero top principal stresses as well as the deflection and maximum (top) stress along the diagonal from the center point to the support corner, like Figure 8.5.

The deflection plot in Figure 8.5 shows that the deflection curvature reverses its sign as it approaches the corner. The corner would lift up, but the *assumed* edge restraint requires that it not move. Therefore, tension forces must develop in the corner reactions to pull it down in the *assumed* restrained position. If the material along the restraint edge is capable of developing a resisting downward force, then you have the correct solution to the actual problem. Otherwise, you have the solution to the *assumed* problem. Unfortunately, many finite element studies give results for the assumed part behavior instead of the actual part behavior. Then, the plots are pretty, but wrong.

If the edges of the plate are simply sitting on top of the walls, then the wall could not pull down on the corner. An air gap would open; the corner would lift up off the wall, and all line reactions would be in compression where the plate remains on the wall. Sometimes you can actually see this corner lift off behavior in thin acoustical ceiling tiles. How much of the corner actually lifts off the wall must be computed from an iterative contact analysis.

If you did not have a contact analysis capability you could still get a reasonable answer to the lift off analysis. To do that you could introduce a split line on each edge near the corner (with parametric dimensions). Let the short end of each corner line be unsupported, solve the problem and check the reactions. If any negative reaction forces appear, then move the split line away from the corner and repeat the process. It may be a slow procedure, but it can lead you to the correct lift off regions.

Shell Analysis

9.1. SolidWorks Shell Capabilities

A general shell is different from a membrane shell, which has only in-plane loads and displacements, and a flat plate shell, which has only transverse loads and in-plane moment vectors. A general shell can have both in-plane and transverse loadings resulting in in-plane force resultants, tangential resultant moment vectors, and transverse resultant shear force vectors. The surface of a shell can be doubly curved (like a sphere or hyperbolic paraboloid), single curved (like a cone or cylinder), or flat. The shell is defined by a mathematical mid-surface and has half the physical thickness on either side of that thickness. Thus, it is thought of as having three physical surfaces (top, mid, and bottom) with different stress levels even though it is displayed as a single surface. The top and bottom sides of the shell are shown in different colors. Shells of different thicknesses can also be shown in different colors.

Within SW Simulation there are three different options for creating shell models: (1) From sheet metal parts having a constant thickness, (2) from surface geometries that have their piecewise constant thickness defined in a study, and (3) from relatively thin solid bodies by extracting their mid-surface or offset surface and defining the associated thickness later in a study session. Table 9.1 lists the current restraint options for mid-surface and offset surface shells within SW Simulation. Fixtures on an edge of a shell are applied directly to the edge. Note that shell load options are given in Table 9.2.

Table 9.1. Fixtures for mid-surface or offset shell stress analysis.

Fixture type	Shell definition
Circular Symmetry	Normal and tangential displacements on repeated surfaces match.
Connectors	See SW Simulation help files for bolts, pins, spot welds, etc.
Fixed Geometry	All translations and rotations are zero on an edge, or vertex.
Fixed Hinge	On a cylindrical face, only the circumferential displacement is allowed.
Immovable	All three translations are zero on a face, edge or vertex.
On Cylindrical Faces	The cylindrical coordinate displacements and rotations normal to and/or on the cylindrical surface are given.
On Flat Faces	Displacements and rotations normal to and/or tangent to the flat face are specified.
On Spherical Faces	The spherical coordinate displacements and rotations normal to and/or on the spherical surface are given.
Roller/Slider	Two displacements tangent to a flat face and the rotation normal to the flat face are allowed.
Symmetry	Select the symmetry plane containing shell edge or vertex to be restrained. (Zero normal displacement and in-plane rotations.)
Use Reference Geometry	A face, edge, or vertex can translate and or rotate a specified amount relative to a reference plane and axis.

When a thin, variable thickness, solid part clearly has a mid-surface description some finite element systems will generate a variable thickness shell elements by interpolating the thickness at each mesh node. SW Simulation currently (2009) does not offer that feature. The mid-surface shell option only works if the two selected surfaces have a constant thickness between them. For piecewise constant thickness solids it is easy to convert them to shell models by using an **Insert → Surface → Extrude** or **Mid-surface** or **Offset-surface** or **Planar-surface** option. Both the solid and the surfaces will continue to exist. When you begin a simulation study

Table 9.2. Load conditions for mid-surface or offset shell stress analysis.

Load type	Shell definition
Apply Force	The total force on a mesh face is specified, or given on a side face or edge to define the mid-surface edge or vertex value.
Apply Normal Force	The total force normal to a face, at its centroid, is specified and converted to an equivalent pressure.
Apply Torque	The total torque on a face is specified with respect to an axis and converted to an equivalent pressure.
Bearing Load	On a cylindrical surface give the total force in a Cartesian X or Y direction to convert to a sine distribution pressure.
Centrifugal	The angular acceleration and angular velocity are given about an axis, edge, or cylindrical surface.
Gravity	The gravitation acceleration value is given and oriented by an axis, edge, or a direction in or normal to a selected plane.
Remote Load	See SW Simulation help files.
Temperature	Not recommended. Transfer from thermal analysis.

you can turn off a solid and retain its shell surface by a right click on **Solid Part Name → Exclude from Analysis**. That is reversible using an **Include in Analysis** pick.

9.2. Quarter Symmetry Tank Stress

You need to carry out the stress analysis of an outdoor water tank. Since it has quarter symmetry you can start by building only one-fourth of the geometry. The bottom is 0.5 inch thick while the walls are 0.25 inch thick. The side wall has height of 72 inches and a small lip extends below the tank bottom for 3 inches. The lower lip will give you more realistic options on how you may need to restrain the part. The complete tank and the dimensions of the tank bottom and the final quarter symmetry part are seen in Figure 9.1. The material is galvanized steel selected to resist rusting.

The tank is to be operated with the water level 6 inches from the top. The tank is analyzed as a shell, so it is constructed as a surface model. The bottom plate is created from the center

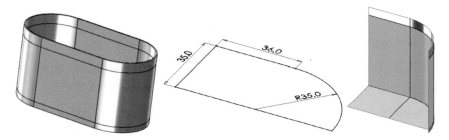

Fig. 9.1.   Tank and its final quarter symmetry region.

sketch using **Insert → Surface → Planar surface**. The top
shell wall is obtained from the side line and arc using **Insert →
Surface → Extrude → Two directions**. The lower support lip is
the second extrude direction. Two horizontal split lines were added
to mark the top and bottom water levels (to aid selection of pressure
faces). A split line was added to the bottom, at the beginning of
the arc, to serve only as a reference direction for restraints. At
this point the **Part name** will appear in the Simulation Manger
menu: Right click on it to apply material data to all of the shells.
Pick **Apply Material to All → Material panel → From
library files → Steel** and select galvanized steel, set the **Units**
to English. Note that the yield strength, taken from a uniaxial
tension test, is about 29.6e3 psi. Since you selected a ductile material,
that material yield property will later be compared to the von
Mises, or effective, stress. Our Safety of Factor (for this material)
will be this yield stress property divided by the von Mises stress.
Click on the part name to show the two shells. Right click on the
wall shell **Edit Definition → Thin → Thickness** and enter
0.25 inch. Repeat for the bottom shell, but use a thickness of
0.5 inch.

Remember that the actual displacement supports (restraints) can
be unclear and you usually need to check for a few possibilities.
What looks like minor changes in the restraints of a part can cause
large changes in the displacements and/or stresses. Also, remember
that in a static analysis you must always provide enough restraints
to prevent all of the six rigid body motions (RBMs) possible in a

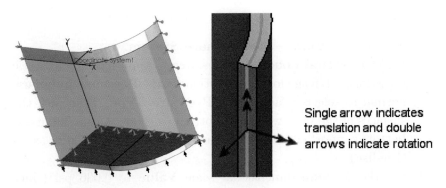

Single arrow indicates translation and double arrows indicate rotation

Fig. 9.2. Tank base support (blue) and symmetry restraints.

three-dimensional part. In this example you will use an initial set of restraints, carry out the analysis, evaluate the study, and add new restraints for an additional analysis. Here, begin by supporting the bottom tank edge against vertical motion (only). That prevents three rigid body motions: motion in the vertical direction and rotation about the two horizontal axes. Eliminate the three RBM that remain by using the two symmetry planes. The symmetry planes are the front $(x$–$z)$ and left $(y$–$z)$ planes. The edges on those planes have their displacement normal to the plane restrained, as well as the shell rotational components lying in the plane. Those restraint sets are assigned different colors (symmetry green), and are shown in Figure 9.2. They were all imposed using **Fixtures** → **Advanced Fixtures** → **Use Reference Geometry**.

In order to apply a variable pressure you first need to create a local coordinate system. Thus, prescribe a hydrostatic pressure load increasing from the top water level marked by the split line:

(1) Go to **Insert** → **Reference Geometry** → **Coordinate System** to open the **Coordinate System panel**.
(2) Locate the **origin** at the point shown in Figure 9.2 and accept the default directions, so that the pressure is a function of the local y-axis. [*WARNING: When a pressure changes signs, SW expects a split line or split surface to be inserted into the model along the zero value contour.*]

Continue with the application of the pressure loading:

(1) Turn on the **View → Coordinate System**.
(2) Select **External Loads → Pressure** to impose the hydrostatic pressure load in the local y-direction. In the **Pressure panel** use normal to selected face as the **Pressure Type**. Then pick the surfaces of the tank walls and bottom (but not the small outside bottom support edges). Select the **Pressure Value Units** as psi (English).
(3) Set the pressure dimensional scale **Value** to 0.036 psi (since the water density, γ, is 0.036 lb/in^3). That value is multiplied times the non-dimensional quadratic polynomial, in the local x–y coordinate directions, activated by checking a **Non-uniform Distribution**. Set all the non-dimensional polynomial coefficients to zero except for the unity **Y** term (so as to create a linear pressure increase with vertical depth).
(4) **Preview** gives you a visual check of the pressure distribution along the edges of the loaded faces.

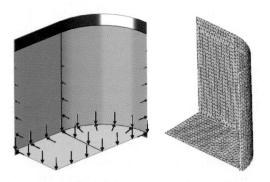

You should expect the highest bending stresses will be near the tank bottom-side wall junction region. Thus, we will eventually probably have to control the mesh to make the smallest elements occur there. However, for the first analysis you can accept the default mesh generation (above).

Now you can right click on the model name and select **Run** to start the first mesh analysis. Passing windows will keep you posted on the number of equations being solved and the status of the displacement solution process and post-processing. You should get a notice that the analysis was completed (not a failed message). Then you have access to the various SW Simulation report and plot options needed to review the first analysis.

Start by double clicking the **Displacement Plot 1** icon. The default plot is a smoothly filled (Gouraud) contour display of the resultant displacement magnitude and the deformed shape part. However, since displacements are vector quantities consider a vector plot first:

(1) Access them from a right click, **Edit Definitions** → **U Resultant** → **Advanced** → **Vector Plot**.
(2) **Edit Definitions** → **Vector Plot Options** double click again on the plot icon to create the view sin Figure 9.3 (left).

Still, the contours values are useful at times. If you do not have a color printer and/or if you want a somewhat finer description you may want to change the default plot styles:

(1) Click in the graphics window and select **Edit Definition** and cancel the advanced option.

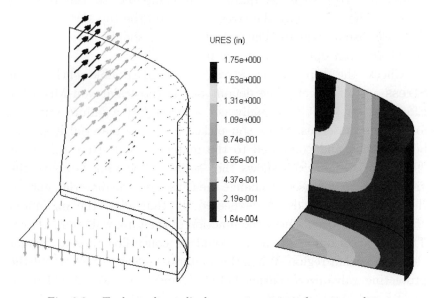

Fig. 9.3.   Tank resultant displacement vector and contour plots.

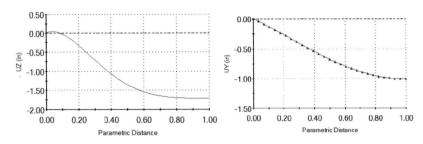

Fig. 9.4.   Maximum wall (left) and bottom deflections.

(2) Double click again on the **Plot** icon to get both the magnified deflected shape and the color contours of the displacement values. That alternate view may be easier to understand or to plot in grayscale.

The detail graphs of the deflection normal to the wall and bottom are given in Figure 9.4. Note that the peak deflection at the top of the tank is several times the thickness of the shell. Thus, this problem definitely will need to be re-run with a large deflection iterative solution.

Recall that a shell element can have different stress levels on the three available surfaces. Here, experience suggests that the membrane stress will be quite small compared to the bending stresses. Therefore, only the stress results on the top surface of the shell are displayed here. The bottom surface should have the same magnitude, but opposite sign (try it).

Check the top surface stress levels by double clicking on **Stress → Plot** icon. The default one is the scalar von Mises (or Effective) stress. It is actually not a stress but a failure criterion for ductile materials. Since you picked a ductile material it should be examined and compared to the material yield stress (of about 29,600 psi). Figure 9.5 shows that some of the tank is above the yield point, so you need to change the thickness, the material, and/or the restraint methods. As expected, in that plot, the maximum effective stress occurs near the junction of the tank wall and bottom. That suggests our next mesh should be controlled to give smaller elements in that region. When this material is near its yield point the protecting galvanized coating fails first and the material will begin to rust and loose strength.

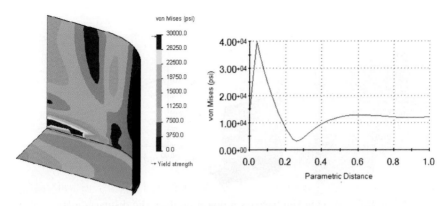

Fig. 9.5. Effective stress distributions, and its graph up the tank wall.

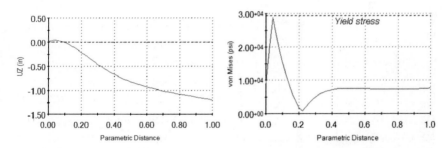

Fig. 9.6. Large deflection result for wall deflection and effective stress.

This part was re-run with the large deflection option. For some problems, that option shows increased membrane stresses that are missed by classical small deflection theory. However, in this case the changes were small. The new graphs are seen in Figure 9.6. The tank will have to be changed to avoid damaging the galvanized coating.

This model could also be revised to look at other restrain conditions. For example, if the tank base sits on two $2'' \times 4''$ wooden boards (at the bottom edge) you should expect higher stresses. That requires additional split lines on the current bottom edge lip surface to pick a smaller support surface. The results in Figures 9.7 and 9.8 show that the large displacements double, but there is little change in the peak stresses.

Likewise, if you assumed that the tank sinks into the ground (or you eliminate the bottom edge) so the full tank bottom is supported

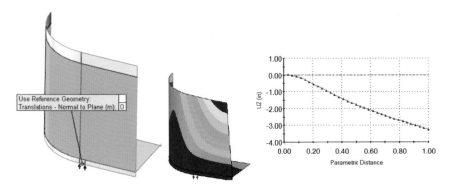

Fig. 9.7.   Placing the base on a board doubles the displacements.

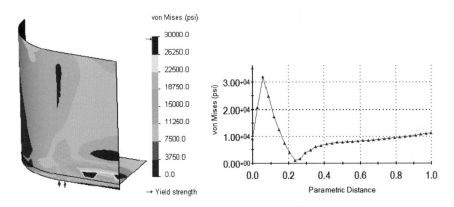

Fig. 9.8.   Effective stresses for a board supported tank.

in the vertical direction then you could use the existing geometry. You would just set a vertical (normal only) restraint there. The stresses and deflections would be much smaller in that case. To test that concept, you would only need to add one additional restraint set that provides vertical support to the tank bottom plate.

9.3. Solid Stress Analysis Approximation with 2.5D

There are many 3D parts that can be represented with a 2D drawing of regions noted as having different constant thicknesses. Components of that sort are commonly referred to as 2.5D solids.

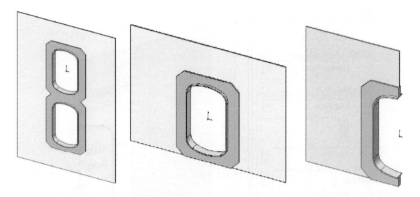

Fig. 9.9.   Portals through a ship bulkhead gangway.

They can be analyzed with shell models, for any loading states, as a way to validate full 3D solid studies and/or to help plan the mesh controls needed to make the initial 3D study economical. As an example, consider a ship bulkhead that is subject to an in-plane constant bi-axial stress state, with $\sigma_x = \sigma$ and $\sigma_y = \sigma/2$. The bulkhead contains two symmetrical portals (Figure 9.9) that are 1 m wide. The openings will cause a local stress concentration, say σ_{max}, at their edge (to be shown below). Since this is a linear analysis, the results can be directly scaled for any value of σ.

The stress concentration factor, K_t, for bi-axial tension around an elliptical hole in an infinite plate is [11]:

$$K_t = 1 + \frac{2a}{b} - \frac{\sigma_y}{\sigma_x} = 1 + \frac{2(0.875)}{0.5} - \frac{\sigma/2}{\sigma} = 4.0,$$

where a and b are the major and minor axes of a similar ellipse. To reduce the stress concentration factor around the opening, the wall thickness is to be increased in two stages. Employ thickness ratios of 1:4:10 relative to the standard thickness of 0.02 m. Note that these wall thicknesses could be parameters in a weight optimization study. The dimensions on the two regions of increased wall thickness are seen in Figure 9.10.

The sketch of the three regions is extruded relative to the mid-plane of the bulkhead, with merge results checked, to form a 3D solid. The quarter symmetry model with loads and restraints and its solid

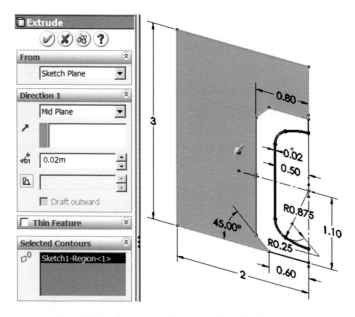

Fig. 9.10. Regions of increased wall thickness.

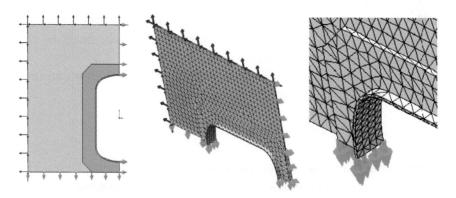

Fig. 9.11. A quarter symmetry solid model and solid mesh.

mesh is given in Figure 9.11. There is one quadratic element through the thickness of the main bulkhead. That is sufficient, since there are no transverse loads to cause bending. Otherwise, mesh control would be required to force more solid elements into the thickness. Before continuing on to the structural solid results, the creation of the 2.5D shell validation model will be introduced.

The original solid was extruded as three merged constant thickness regions, about a common mid-plane. Several commercial finite element systems could mesh such a solid with mid-surface shell elements and automatically assign the correct thickness to each element. SW Simulation currently (2009) does not do that. You can however employ an assembly of the three regions, bonded together, each consisting of a mid-surface membrane shell having a specified thickness. Here, you need at least two of those three shell surfaces to be in the same plane.

Extrude the first region just like it is shown in Figure 9.10. Save that body, named as "Thin", and suppress it so it does not merge with the next extrusion. Extrude the next body with the same thickness, name it "Mid" and suppress it. Extrude the third region with the same thickness and save it with the name "Thick". Import the three bodies into an assembly and mate them together. As shown in Figure 9.12, you can set each body to have a different thickness.

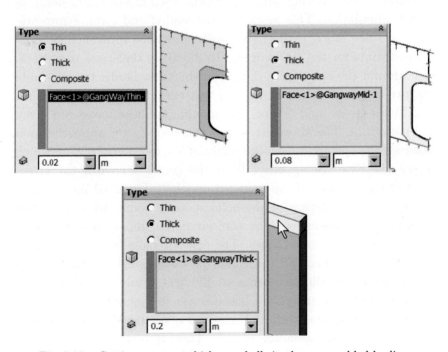

Fig. 9.12. Setting constant thickness shells in three assembled bodies.

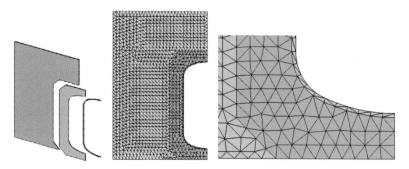

Fig. 9.13.   Flat shell, three thickness, bounded body assembly and mesh.

The last two regions were defined as thick, although that was probably not necessary for this in-plane loading state. If the edge of the opening had been much thicker it could have also been re-run as an out of plane shell as another validation bound estimate.

Figure 9.13 shows the three imported bodies in the SolidWorks assembly, before mating, and the created shell meshes after bonding in SW Simulation. This assembly was loaded and given symmetric restraints like in Figure 9.11 (but with additional rotational symmetry restraints for the shell edges). To illustrate the stress concentration around the opening of the single thickness model (which omits the two thicker reinforced regions) was executed. Its displacement, von Mises stress, and (twice) the maximum shear stress are given in Figure 9.14. The above stress concentration factor approximation assumed the opening was in the center of a symmetrical region, which is not the case. Being offset from the center increases the stress level at the bottom of the modeled opening. The 2.5D model gives a very good validation of the solid model results, with a lot less computational resources. The contour plot comparisons are set to have the same contour ranges. The contour plots have the solid on the left of the figure and the 2.5D (piecewise constant shell) model on the right. The displacements are illustrated in Figure 9.15.

The von Mises stress comparison is seen in Figure 9.16. Compared to the non-reinforced wall of Figure 9.14, the peak stress has been reduced by about a factor of 7.5. That is seen more clearly in Figure 9.17 which gives the graph of the von Mises stress along the

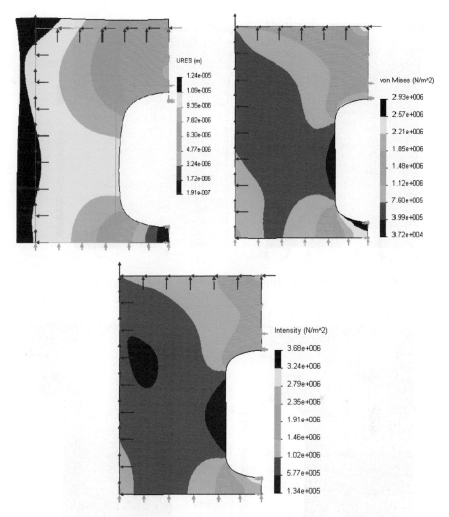

Fig. 9.14.   Constant thickness bulkhead: displacement, von Mises, intensity.

vertical line from the bottom of the opening to the bottom of the model. The intensity (twice the maximum shear stress) from the solid and 2.5D models are given in Figure 9.18. The validations are good.

The above three thickness shell model did not catch some of the 3D response of the material adjacent to the hole. The flanges of the curved region around the hole of the model did not have constant displacements. The mid-plane moved the most, while the outer edges

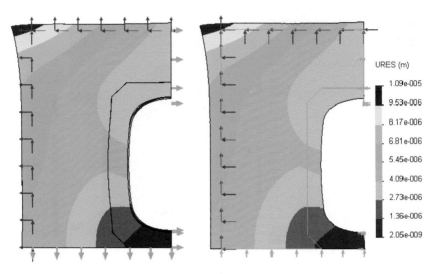

Fig. 9.15.   Bulkhead displacement results: solid (left), 2.5D (right).

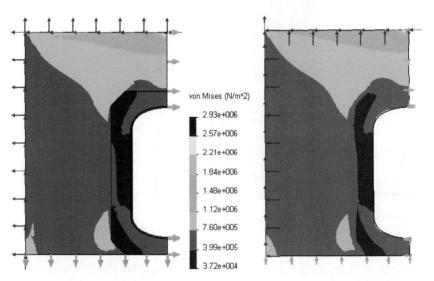

Fig. 9.16.   Bulkhead von Mises stress results: solid (left), 2.5D (right).

of the flange were seen to move less. That is, there was a relative, symmetric, slight curving (bending) into the opening by the thickest region. A graph of the flange bending displacement, from front to back, is given in Figure 9.19. The current 2.5D model missed that

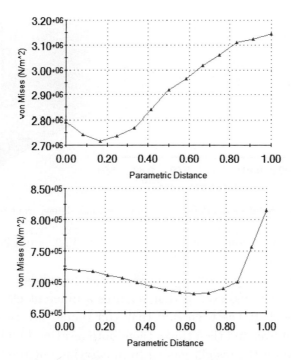

Fig. 9.17. Gangway stress reduction from single (top) to three thicknesses.

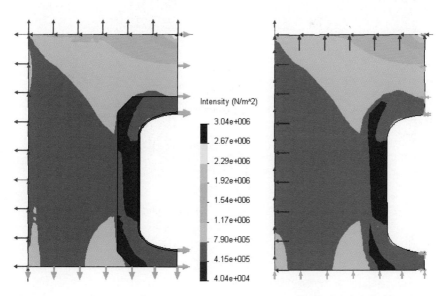

Fig. 9.18. Twice the maximum shear stress in the bulkhead: solid (left), 2.5D.

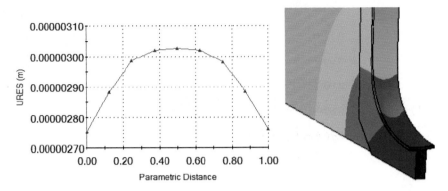

Fig. 9.19.   Solid model deflection along the thick flange, front to back.

very small feature, but an out of plane shell model would have shown a similar result.

It is always important to consider ways to validate your finite element calculations, even if that requires a different class of finite element model. It has been said that you should use two different models, as above, and then throw them both away and build a better model based on the insight gained from carrying out the first study and its validation.

Space Truss and Space Frame Analysis

10.1. Introduction

One-dimensional models can be very accurate and very cost effective in the proper applications. For example, a hollow tube may require many thousands of solid elements to match its geometry, even though you expect its stresses to be constant. A truss (bar) or frame (beam) element can account for the geometry exactly and give "exact" stress results and deflections with just a handful of equations to solve. In SW Simulation, truss and frame elements are available only for static (constant acceleration), natural frequency, buckling, or nonlinear studies. It is recommended that you review the SolidWorks weldments tutorial and the truss tutorial before using trusses or frames.

The truss element is a very common structural member. A truss element is a "two force member". That is, it is loaded by two equal and opposite collinear forces. These two forces act along the line through the two connection points of the member. The connection points (nodes) between elements form a concurrent force system. That is, the joints transmit only forces. Moments are not present at the joints of a truss. The truss elements in SW Simulation are all space truss elements. There are three displacement DOF at each node (see the right side of Figure 3.1), and up to three reaction forces at a restrained joint. A space truss has six rigid body motions, all of which must be restrained in an analysis. The space truss and space frame models are created in SW Simulation by 3D line sketches. For a truss the lines must exactly meet at common points (joints). The lines of the space frame models can meet at common points and/or terminate as an intersection of two lines. To avoid numerical ill-conditioning, it is best if a space frame does not have two joints very close (say, the width of the cross-section) to each other. If that is necessary, SW Simulation takes special action to build the finite element model there.

The line that represents a truss or frame member has to be located relative to the cross-section of the member. Where it intersects the cross-section is called the pierce point. For trusses it is important that the pierce point be at the centroid of the cross-section. That happens automatically when you use a built in library shape. If you construct a cross-section make sure that the truss pierce point is at the centroid. If the pierce point is not at the centroid, as in the right of Figure 10.1, then an axial load will cause bending stresses to develop and to be superimposed on the axial stress. That is allowed in frame elements but not truss elements.

Clearly, the elastic bar is a special form of a truss member. To extend the stiffness matrix of a bar to include trusses in two- or three-dimensions basically requires some analytic geometry. Consider

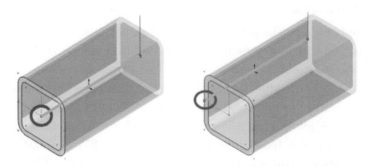

Fig. 10.1.   Centroidal (truss) and eccentric (beam) section pierce points.

a space truss segment in global space going from point 1 at (x_1, y_1, z_1) to point 2 at (x_2, y_2, z_2). The length of the element between the two points has components parallel to the axes of $L_x = x_2 - x_1$, $L_y = y_2 - y_1$, $L_z = z_2 - z_1$ and the total length is $L^2 = L_x^2 + L_y^2 + L_z^2$. The direction cosines are defined as the ratio of the component length increments divided by the total length of the element, L. They are used to transform a bar stiffness matrix to the space truss stiffness matrix. For 2D problems, only one angle is required to describe the member direction. A truss element stiffness requires only the material elastic modulus, E, the cross-sectional area, A, and the member length, L. A space frame element also requires the three geometric moments of inertia of the cross-section. Two inertias are needed for the transverse bending, and the third is needed for torsional effects. The SW Simulation frame element also utilizes the material's Poisson's ratio. The mass density, ρ, is needed for gravity (acceleration) loads, or natural frequency computations. Weight loads acting on a truss are transferred to its two end nodes as forces only, and the mid-span bending effect of the weight is ignored.

If you combine the bar member, which carries only loads parallel to its axis, and a beam which carries only loads transverse to its axis you get the so-called beam-column element, or general frame element. When deflections are large, the iterative solution updates the axial load in a beam and that can significantly affect the results. Adding the ability to carry torsion moments along the element extends the behavior to a space frame. In other words, a space frame is a combination of individual beam-column elements that resists loadings by a combination of transverse bending moments, axial member forces, and transverse (shear) forces, and an axial torsional moment. Therefore, it is a more efficient structure than a space truss element. Weight loads acting on a frame are transferred to its two end nodes as both forces and couples. Therefore, the effect of the weight is included in the deflection and stresses along the full length of a frame element.

10.2. Statically Determinate Space Truss

Consider the simple symmetric space truss shown in Figure 10.2. It has two horizontal members, denoted by a, and an inclined member, b, which is located in the vertical mid-plane. The truss has three immovable restraints (at the dashed circles) and a vertical point load, $P = 1,000$ lb at the free node. The spatial dimensions of the nodes are shown in the figure. The members are square hollow tubes, with the horizontal pair being $2 \times 2 \times 0.25$ inches, and the other $4 \times 4 \times 0.25$ inches. All three members are made of ASTM A36 steel.

The construction of the 3D line models is done by means of a **3D Sketch**:

(1) **Insert** → **3D Sketch**. Insert construction lines along each of the axes to help locate nodes (remember to press Tab to change to a new plane). Add and dimension (in inches) additional construction lines in each coordinate plane. In the **3D Sketch** → **Lines**. Draw one line to open the **Line Properties panel**.
(2) Expand the **Additional Parameters** option to provide access to the end points of the first element.
(3) Specify the coordinates of the **starting point** and/or the **ending point** and/or the **increments** in coordinates from one

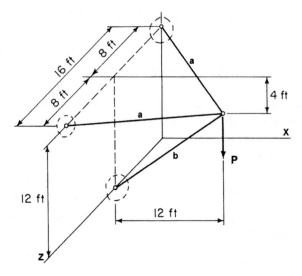

Fig. 10.2.   Geometry of the simplest space truss.

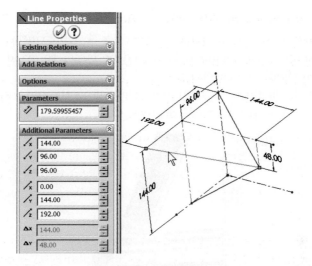

end. Spot check the element **length** value. Click **OK**. Repeat for the other elements (or review the 3D sketch tutorial for other approaches).

After all the space truss line elements have been located, and saved, the next task is to look up or construct each member's cross-section:

(1) **View → Toolbars → Weldments** ▨ and click **Structural Member** ▨

(2) In the **Structural Member panel, Selections → Type → Square tube**. Pull down to the $2 \times 2 \times 0.25$ inches size. Select the two horizontal elements as the **path segments**. Under **Settings → Apply corner treatment** and click on **end miter** and **OK**.

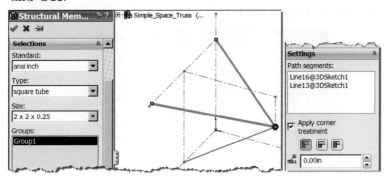

(3) In the **Structural Member panel Selections** → **Type** → **Square tube**. Pull down to the $4 \times 4 \times 0.25$ inches size for the compression member. It should be larger to avoid buckling (which should be checked).

(4) Select the inclined, vertical plane element as the **path segment**. Change **Alignment** to 45 degrees, **OK**.

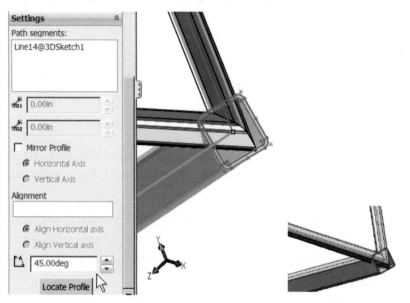

In the SW Simulation menu:

(1) Right click on **Simulation** → **New Study** → **Static**.
(2) Select all three beams, **Edit definition** → **Apply/Edit Beam** → **Truss**. Click **OK**.

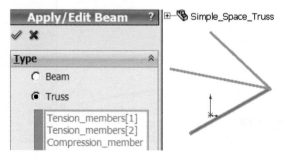

(3) Right click on **Truss name** → **Apply Material to All Bodies** → **Library** → **Steel** → **ASTM A36**, click **OK**.

(4) Right click on **Joint Group** → **Edit** → **All** → **Calculate**. SW Simulation verifies and displays all four joints. You can manually pick joints also.

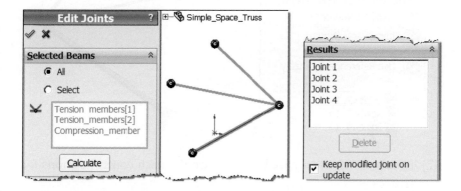

(5) **External Loads** → **Force**. In the **Force panel** → **Select Joints**. Pick the one free node and select the **Front plane** (click + sign on part name tree, pick Front Plane) to set the plane. Select the desired tangential (vertical) direction. Select lb **units**, and set a **value** of 1,000 lb as the vertical component. Click **OK**.

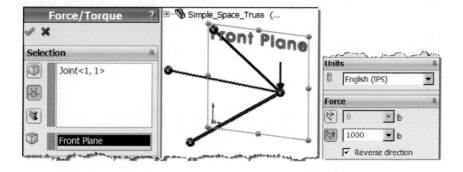

(6) **Fixtures** → **Fixed Geometry** → **Immovable**, select the three wall ball joints. They prevent the three rigid body translations and three rotations. Click **OK**.

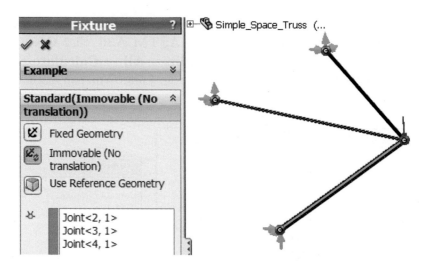

(7) **Mesh → Create Mesh**. There are no mesh control options for trusses.

(8) **Run**. SW Simulation calculates the space truss joint locations and their displacements.

In the Simulation Menu: Right click **Results → Define Displacement Plot → URES: Resultant displacement**. The deformed plot and the probed maximum displacement are in Figure 10.3.

In the Simulation Menu: Right click **Results → Define Displacement Plot → Resultant Reaction Force**. Set the **units**

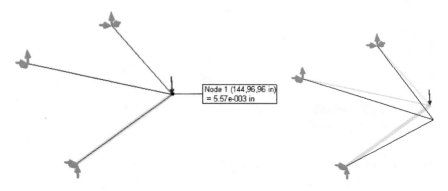

Fig. 10.3. Space truss scaled deflection.

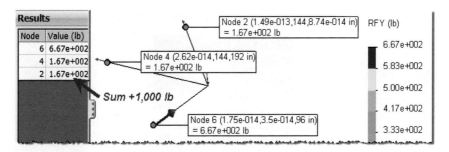

Fig. 10.4. Space truss vertical reaction force probes and sums.

to lb and select **Advanced → Vector Plot, OK.** Use **Vector Plot Options** to control the vector lengths. Repeat the process for the vertical reaction components (**RFY: Y Reaction Force**) and verify that they sum to a value that is equal and opposite (+1,000 lb) to the total applied vertical load(s). It is also wise to verify that both the x- and z-components of the reaction forces sum to zero. These two reaction probe sets are illustrated in Figure 10.4.

In a truss, you are usually interested in both the member forces and their axial stresses:

(1) To see the list of member forces use **Results → List Beam Forces → Forces** to select the **units** and the **Beam Range** numbers of interest. You can save those data as comma separated value files.

List Beam Forces ?
✔ ✘ ↩

List ☆
⦿ Forces
○ Stresses
🄵 English (IPS) ▼

Beam Range ☆
Start: 1
End: 3

Beam Name	Element	End	Axial
Beam-1(Tension_members[1])			
	1	1	-623.61
		2	623.61
Beam-2(Tension_members[2])			
	2	1	-623.61
		2	623.61
Beam-3(Compression_member)			
	3	1	1201.9
		2	-1201.9

(2) In a similar manner you can list the axial stress in truss members.

In the Simulation Menu: Double click on **Stress 1** to show the default plot. Select different view points, as in Figure 10.5. Use **Edit Definition** → **Stress Plot** and select **Axial** and set the **units** to psi. Note that each truss member, by definition, has a constant axial stress level. The member stress levels are very low compared to the material yield point. It is unlikely that a buckling study of the compression member is needed, even if you included the neglected weight of the members.

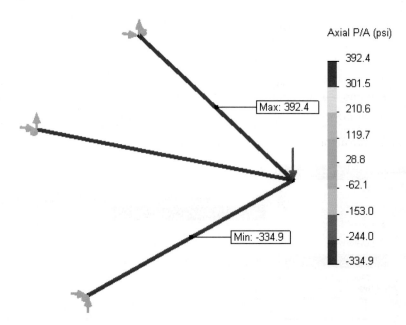

Fig. 10.5.   Truss member's constant axial stress.

10.3. Analytic Beam Element Matrices

An elastic beam acts like a generalized spring with two nodal degrees of freedom at each end: the transverse displacement, v, and the slope, θ. However, in addition to end point transverse shear load, V, and moment, M, it can have distributed transverse loads per unit length, $q(x)$, and/or thermal moment due to a temperature change, say ΔT, through its thickness, h, from bottom to the top. The resultants of such effects are lumped at the ends as additional point forces and/or moments. For a beam with a cross-sectional moment of inertia, I, length, L, and material with an elastic modulus of E and a coefficient of thermal expansion of α, the corresponding analytic 4 by 4 element equilibrium matrices are:

$$
\frac{EI}{L^3}
\begin{bmatrix}
12 & 6L & -12 & 6L \\
6L & 4L^2 & 6L & 2L^2 \\
-12 & -6L & 12 & -6L \\
6L & 2L^2 & -6L & 4L^2
\end{bmatrix}
\begin{Bmatrix}
v_1 \\ \theta_1 \\ v_2 \\ \theta_2
\end{Bmatrix}
$$

$$
=
\begin{Bmatrix}
V_1 \\ M_1 \\ V_2 \\ M_2
\end{Bmatrix}
+ \frac{L}{60}
\begin{bmatrix}
21 & 9 \\
3L & 2L \\
9 & 21 \\
-2L & -3L
\end{bmatrix}
\begin{Bmatrix}
q_1 \\ q_2
\end{Bmatrix}
+ \frac{\alpha \Delta T E I}{h}
\begin{Bmatrix}
0 \\ 1 \\ 0 \\ -1
\end{Bmatrix},
$$

where q_1 and q_2 are the transverse distributed load per unit length at the first (left) and second end, respectively. The point sources V_k and M_k represent externally applied force and moment at end k. In matrix notation the equilibrium equation is

$$
[\boldsymbol{K}^e]\{v\} = \{\boldsymbol{F}_p\} + \{\boldsymbol{F}_q\} + \{\boldsymbol{F}_T\},
$$

which equates the product of the elastic stiffness and the generalized displacements to the sum of the generalized forces resulting from point sources, distributed sources and thermal sources.

This specific element is called the cubic beam because the four generalized nodal displacements define a cubic polynomial in one-dimension. In other words, the transverse deflection of the beam, $v(x)$, varies cubically with the position along the axis of the beam, x. This is one of the most widely utilized finite elements. For any of the load combinations given on the right side of the above matrix expression the solution gives the analytically exact results at the nodes. However, the results (deflection, slope, moment and transverse shear) interior to the element are only exact when the distributed load is constant ($q_1 = q_2$). Otherwise, you need a fine mesh of beam elements to recover accurate transverse shear results.

In SolidWorks notation, the x-direction above is called the axial direction and the V and M results are computed with respect to local beam direction-1. A similar set of matrix relations occur in the orthogonal plane, called direction-2, when utilized in three-dimensions as a space frame.

The analytic beam equations can often be used to give simple closed form validation estimate for real problems that must be obtained by numerical finite element studies. As the first example, consider finding the deflections and reactions of cantilever beam with a triangular distributed line load shown in Figure 10.6.

The resultant end forces and moments from the line load become:

$$
\mathbf{F}_q = \frac{L}{60}
\begin{bmatrix}
21 & 9 \\
3L & 2L \\
9 & 21 \\
-2L & -3L
\end{bmatrix}
\begin{Bmatrix} 0 \\ w \end{Bmatrix}
= \frac{wL}{60}
\begin{Bmatrix} 9 \\ 2L \\ 21 \\ -3L \end{Bmatrix}.
$$

The first (left) end is free, so the shear force and moment there vanish, $V_1 = 0$, $M_1 = 0$. The unknown wall reaction shear and moment acts on the second node. The resultant load $R = wL/2$ acts at a third of the distance from the wall.

The essential boundary conditions are that the right end deflection and slope are zero. They leave only the first two rows of the beam equilibrium equation for finding the tip deflections:

$$
\frac{EI}{L^3}
\begin{bmatrix} 12 & 6L \\ 6L & 4L^2 \end{bmatrix}
\begin{Bmatrix} v_1 \\ \theta_1 \end{Bmatrix}
= \begin{Bmatrix} 0 \\ 0 \end{Bmatrix}
+ \frac{wL}{60} \begin{Bmatrix} 9 \\ 2L \end{Bmatrix}
+ \begin{Bmatrix} 0 \\ 0 \end{Bmatrix}.
$$

Inverting the square 2 by 2 matrix gives the free end deflections as:

$$
\begin{Bmatrix} v_1 \\ \theta_1 \end{Bmatrix}
= \frac{L^3}{12EIL^2}
\begin{bmatrix} 4L^2 & -6L \\ -6L & 12 \end{bmatrix}
\frac{wL}{60} \begin{Bmatrix} 9 \\ 2L \end{Bmatrix}
= \frac{wL^3}{120} \begin{Bmatrix} 4L \\ -5 \end{Bmatrix}.
$$

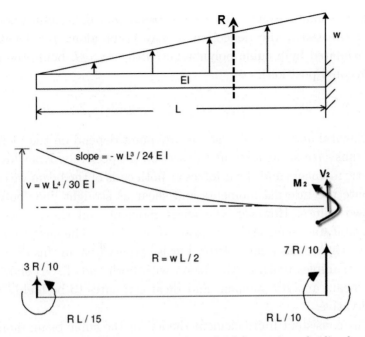

Fig. 10.6.　Beam deflection and resultant load for a triangular line load.

These are the analytically exact end deflection and end slope.

The known exact solution to this problem is a fifth degree polynomial deflection, namely

$$120\,EI\,v(x) = wL^4 \left[4 - \frac{5x}{L} + \left(\frac{x}{L} \right)^5 \right].$$

That means the exact moment in the member $(EI\,v'')$ will be a cubic polynomial, while the transverse shear force $(EI\,v''')$ will be a quadratic polynomial. Nevertheless, the cubic beam element approximation gives exact end moment and shear. The system reactions come from the last two rows of the beam equilibrium matrices:

$$\frac{EI}{L^3} \begin{bmatrix} -12 & -6L \\ 6L & 4L^2 \end{bmatrix} \frac{wL^3}{120} \begin{Bmatrix} 4L \\ -5 \end{Bmatrix} = \begin{Bmatrix} V_2 \\ M_2 \end{Bmatrix} + \frac{wL}{60} \begin{Bmatrix} 21 \\ -3L \end{Bmatrix} + \begin{Bmatrix} 0 \\ 0 \end{Bmatrix},$$

$$\begin{Bmatrix} V_2 \\ M_2 \end{Bmatrix} = \frac{wL}{6} \begin{Bmatrix} -3 \\ L \end{Bmatrix},$$

which are again exact. Since each beam has only four undetermined constants, the deflection of any point along the length is approximated by a cubic polynomial (with $r = x/L$ being the non-dimensional position):

$$v(x) = v_1(1-3r^2+2r^3)+\theta_1(r-2r^2+r^3)L+v_2(3r^2-2r^3)+\theta_2(r^3-r^2)L.$$

The flexural and shear stresses at any point depend on the moment and transverse shear force at the section, respectively. Knowing the analytic moments and shear forces at both ends of the beam we could estimate the internal moments and shear as straight lines between the two values. However, the exact moment and shear diagrams have cubic and quadratic variations, respectively. The analytic cubic beam estimate has a large error, but is conservative in this example. Other examples (such as this beam with both ends fixed) show the cubic beam analytic moment and shear estimates to be highly non-conservative.

The consistent finite element theory for the cubic beam moment and shear are linear and constant along the beam length, respectively. Specifically, they are:

$$M(x) = \frac{v_1(12r-6) + \theta_1(6r-4)L + v_2(6-12r) + \theta_2(6r-2)L}{L^2},$$

$$V(x) = \frac{v_1(12) + \theta_1(6)L + v_2(-12) + \theta_2(6)L}{L^3}.$$

Of course numerical solutions with many elements give very good results. Here we are emphasizing quick analytic estimates based on a mixture of solid mechanics theory and finite element theory. If you used two analytic beam elements you would get much better moment and shear estimates, but you have to solve more equations (try it).

Probably the most common cantilever beam concept is where it is only loaded by a single end force, V_1. Again solving only the last two equilibrium rows:

$$\begin{Bmatrix} v_1 \\ \theta_1 \end{Bmatrix} = \frac{L^3}{12EIL^2} \begin{bmatrix} 4L^2 & -6L \\ -6L & 12 \end{bmatrix} \begin{Bmatrix} V_1 \\ 0 \end{Bmatrix} = \frac{V_1 L^2}{12EI} \begin{Bmatrix} 4L \\ -6 \end{Bmatrix},$$

which gives the classic tip deflection of $V_1 L^2/3EI$.

As a final analytic beam example, consider the cantilever with only a temperature change (cooler on top) that is constant along its full length. For a thermal loading only the right hand side changes to:

$$\frac{EI}{L^3}\begin{bmatrix} 12 & 6L & 4L^2 & 6L \\ 6L & 4L^2 & -6L & 12 \\ -12 & -6L & 12 & -6L \\ 6L & 2L^2 & -6L & 4L^2 \end{bmatrix} \begin{Bmatrix} v_1 \\ \theta_1 \\ v_2 \\ \theta_2 \end{Bmatrix}$$

$$= \begin{Bmatrix} 0 \\ 0 \\ V_2 \\ M_2 \end{Bmatrix} + \begin{Bmatrix} 0 \\ 0 \\ 0 \\ 0 \end{Bmatrix} + \frac{\alpha \Delta T EI}{h}\begin{Bmatrix} 0 \\ 1 \\ 0 \\ -1 \end{Bmatrix}.$$

Since the right end deflection and slope are zero the independent displacement relations (for a negative temperature change) are obtained from the top two rows of the equilibrium equations:

$$\frac{EI}{L^3}\begin{bmatrix} 12 & 6L \\ 6L & 4L^2 \end{bmatrix}\begin{Bmatrix} v_1 \\ \theta_1 \end{Bmatrix} = \frac{-\alpha \Delta T EI}{h}\begin{Bmatrix} 0 \\ 1 \end{Bmatrix}$$

and the free tip deflections (similar to the middle of Figure 10.6) are

$$\begin{Bmatrix} v_1 \\ \theta_1 \end{Bmatrix} = \frac{\alpha \Delta T L}{12h}\begin{Bmatrix} 6L \\ -12 \end{Bmatrix},$$

which results in wall reactions of

$$\begin{Bmatrix} V_2 \\ M_2 \end{Bmatrix} = \frac{\alpha \Delta T EI}{h}\begin{Bmatrix} 0 \\ 0 \end{Bmatrix}.$$

This means that since the beam was free to expand, there is no thermally introduced wall reaction shear or moment. That will not be true for a statically indeterminate beam. For example, had the beam been fixed-fixed the two end reaction moments would be $\mp\alpha\Delta T EI/h$. That is, the beam would have a constant moment and corresponding flexural stresses, but no transverse shear.

10.4. Frame Elements

When you have a member that carries both axial loads, like the bar, and transverse loads, like the beam, the resulting member is known

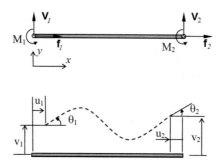

Fig. 10.7.   Combining the bar and cubic beam defines a frame element.

as a frame element or beam-column. A frame element is shown in
Figure 10.7. For planar frame elements there are three generalized
displacements per node (two displacements and one rotation). When
used as a space frame element that number increases to three
displacements and three rotations per node. The third rotation comes
from including torsional effects about the axial direction.

For small deflections, the analytic models given above for the
axial and transverse effects are assumed uncoupled and you just solve
both models independently. Note that for small deflection studies the
axial displacements must be obtained first in order to calculate the
axial force in the bar. If the displaced form rotates the bar then
the axial force has a component in the original transverse direction
and the beam-column becomes coupled. Mathematically, the nature
of the problem changes because the governing differential equation
changes from the classic beam

$$EI\frac{d^4v}{dx^4} = w(x)$$

to a coupled equilibrium

$$EI\frac{d^4v}{dx^4} - f\frac{d^2v}{dx^2} = w(x),$$

where f is a pre-existing axial force (tension positive). The classic equation yields solutions for the beam displacements that are
polynomials. The coupled system requires the exact solution to be
hyperbolic functions instead of polynomials. Of course, the coupled

system can be approximated in FEA studies by using the cubic beam model, but the element lengths need to be greatly reduced. When the cubic beam is used with the pre-existing force it has an additional stiffness term, called the geometric stiffness, which must be added to the matrix equilibrium equations. That matrix is

$$[\boldsymbol{K}^G] = \frac{f}{30L} \begin{bmatrix} 36 & 3L & -36 & 3L \\ 3L & 4L^2 & -3L & -L^2 \\ -36 & -3L & 36 & -3L \\ 3L & -L^2 & -3L & 4L^2 \end{bmatrix}.$$

Notice that a tension force (positive f) increases the stiffness of the beam and thus reduces the transverse deflection $v(x)$. A compression load has the opposite effects. The matrix equilibrium equation becomes:

$$[\boldsymbol{K}^e + \boldsymbol{K}^G]\{v\} = \{\boldsymbol{F}_p\} + \{\boldsymbol{F}_q\} + \{\boldsymbol{F}_T\}.$$

To solve this system you usually employ a large displacement iteration. First, the geometric stiffness is neglected because f is not known. From the first set of displacements the axial bar forces are found as reactions. They are passed on to the beam so its geometric stiffness can be found. The revised system is solved, including $\boldsymbol{K}^G$, to yield new displacements. The process iterates until the changes in the forces and the displacements are very small.

To extend the stiffness matrix of a frame element to include two- or three-dimensions basically requires the use of its direction cosines. The required transformation matrix is similar to the one needed for trusses but with additional terms for the nodal rotations. When such an element is used in general three-dimensional space it is called a space frame element. In that case bending moments can occur in two planes orthogonal to the axis of the element. A space frame also has a torsional element model of the axis as a shaft.

10.5. Statically Indeterminate Space Frame

Consider an unequal leg planar frame that is to be subject to a transverse force and an in-plane support settlement. The cross-section is an ISO $80 \times 80 \times 5$ mm square tube. The tall vertical leg is 15 ft, the short one 10 ft and the top member 12 ft long. A survey shows that the support for the shorter leg has settled and imposed a vertical downward displacement of 0.45 inch and a clockwise rotation of 0.02 radians normal to the plane of the frame. Before considering the transverse load, a study needs to be run to establish the deflections and stresses introduced by the non-rigid support displacement. There are numerous tabulated frame results given in Ref. [11], but none match this load case.

Begin by drawing and dimensioning the three lines defining the frame. Then form the structural members:

(1) **View** $\rightarrow$ **Toolbars** $\rightarrow$ **Weldments** and click **Structural Member**.

(2) In the **Structural Member panel, Selections** $\rightarrow$ **Standards** $\rightarrow$ **ISO**. In **Selections** $\rightarrow$ **Type** $\rightarrow$ **Square tube**. Pull down to the $80 \times 80 \times 5$ mm size. Select all three line segments as the **path segments**.

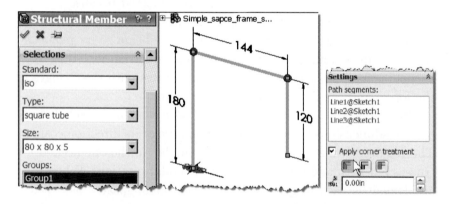

(3) Under **Settings** $\rightarrow$ **Apply corner treatment** and click on **end miter** (above right) and **OK**.

In the SW Simulation menu:

(1) Right click on the **Simulation** $\rightarrow$ **New Study** $\rightarrow$ **Static**.

(2) In the new study, highlight all three beam members, and right click on **Apply/Edit Beam**.

(3) Set the element **Type** to **Beam** and default to **Rigid End Connection** at the ends of all elements. All three space frame members appear now in the beams list, **OK**.

(4) Right click on **Part name** → **Apply Material to All Bodies** → **Library files** → **Steel** → **ASTM A36**, click **OK**.

(5) Right click on **Joint group** → **Edit** → **All** → **Calculate**. SW Simulation locates and displays all four joints (you can manually pick joints as well).

(6) **Fixtures** → **Fixed Geometry**. Pick the lower left support as an encastre (fixed support). This prevents all three rigid body translations and three rotations. Click **OK**. Change the restraint name (in the SW manager menu) to Fixed_lower_joint via a slow double click.

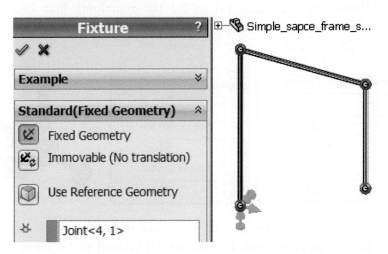

(7) Specify the non-zero support settlements. **Fixtures** → **Use Reference Geometry**. Pick the lower right joint, selected the **Front Plane** as the reference geometry (from the SolidWorks menu, or by expanding the part tree in the graphics area). Set the **Translations** as upward 0.45 inch, and the **Rotation** normal to the front plane (z-axis) as 0.02 radians counterclockwise. Green symbols appear at the two settlements. **Name** the restraint to Settlement.

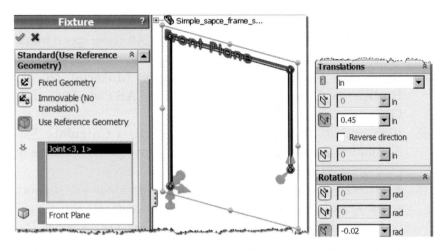

(8) **Fixtures → Use Reference Geometry**. Select the **Front Plane**. Set the zero values for the remaining four DOF at that joint, change their **color** to blue. **Name** the restraint to No_Settlement.

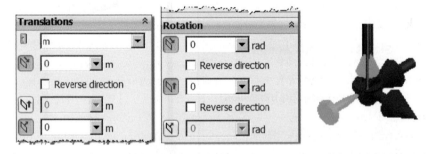

(9) **Mesh → Create Mesh**. There are no mesh control options for beams or frames. The additional created nodes allow more displacement vector displays, and better varying member stress displays.

(10) **Run**. SW Simulation calculates the space frame joint locations and computes their displacements.

Since there were no externally applied forces (just non-zero support displacements) you should expect that the reaction forces to be equal and opposite. Likewise, since the two supports are off-set vertically, the moment reactions should not be equal and opposite, as seen on

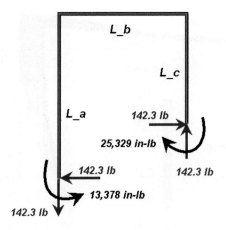

Fig. 10.8.   Frame settlement moment reactions, and system equilibrium.

the left in Figure 10.8. That figure also shows a sketch of the overall reactions, due to the support settlement alone. An FEA always yields the correct reactions for the imposed displacements and loads (none here). You can spot check the system equilibrium, for example, by taking z-axis moments about the right support point: $0 = M_a - M_c - F_x(L_a - L_c) + F_y L_b$ which is satisfied to within 2 in-lb out of at total of 25,330 in-lb. In this case all the reaction loads are in the x–y plane of the frame and the moment vectors are perpendicular to that plane.

In the Simulation menu select **Results → Define Displacement Plot → URES → Advanced → Vector Plots**. Set the display units to inches. In this case, all translational displacements are in the plane of the frame. Figure 10.9 shows that the maximum resultant displacement was about 0.99 inches. The locations of the probed maximum horizontal and vertical displacements are different, as seen in Figure 10.10.

SW Simulation can display the member forces and moments, including the axial force, transverse shear force in two directions normal to the frame axis, two local bending moments, and any torsional moment. For frame members you can have axial stress, torsional stress, transverse shear stress (two sets), and two flexural stresses combined. Therefore, the worst stress at a section varies over the cross-sectional area. The typical combination of stresses, in

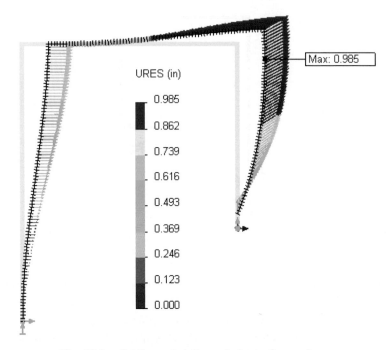

Fig. 10.9.   Settlement deformed planar frame shape.

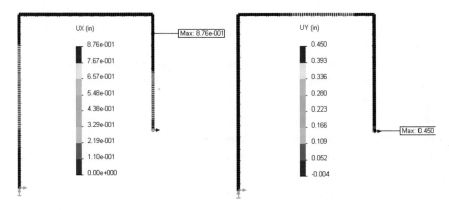

Fig. 10.10.   Maximum x- and y-settlement (right) induced displacements.

one direction normal to the frame, is sketched in Figure 10.11. The
worst stress will usually also vary over the length of each element.
Each frame element is divided into about 50 or more sub-elements
for the purpose of determining, and graphing, the variation of the

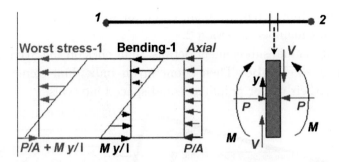

Fig. 10.11. Combining axial stress and flexural stress for the worst stress.

deflections, forces, moments, and stresses along the length of a typical frame element. At the centroid of each sub-element, SW Simulation computes the combined stress at four (or more) locations and displays the worst.

To graph the worst stress value in the frame, utilize:

(1) **Results → Define Stress Plot.** In the **Stress Plot panel**, right click **Edit Definition → Worst case**, set **units** to psi, turn on **Deformed Shape** option, if desired. Click **OK**. Review the default plot.

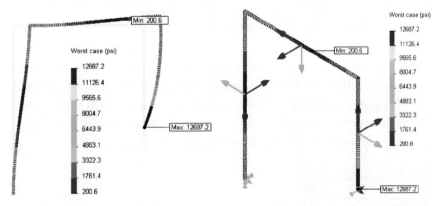

(2) For frame elements it is useful to know the local coordinate directions for each member. They are employed in displaying the internal shear and moment values. To see those local directions right click on **Beam Diagram name → Settings → Options → Show beam directions** (above). The triad

color coding is red for axial, green for member direction 1, and blue for member direction 2.

(3) For individual sub-element details, right click on the **Stress Plot name** → **Probe**. Then zoom in on individual elements and select their center points at equal spaced intervals.

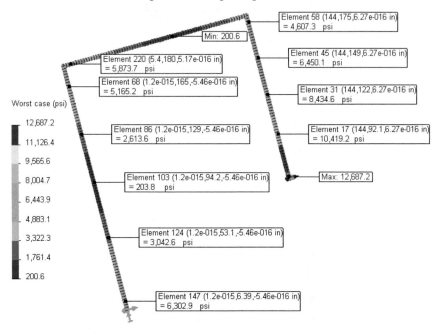

(4) To graph the forces and moments along the frame members use **Results** → **Define Beam Diagrams** to open the **Beam Diagram panel** and there you pick the desired **units**, the **force** or **moment** required, as well as the **Selected Beams** for display. The moment in the top frame member is displayed:

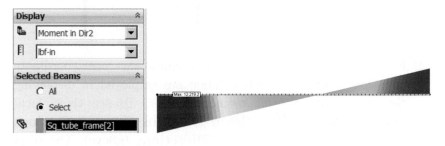

(5) Display the **axial force** (tension positive) you see results consistent with the system FBD in Figure 10.8:

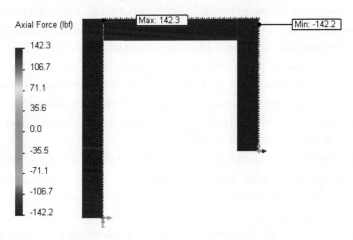

(6) The z-moment, for the first two elements, in direction 2 is likewise displayed (with clockwise positive):

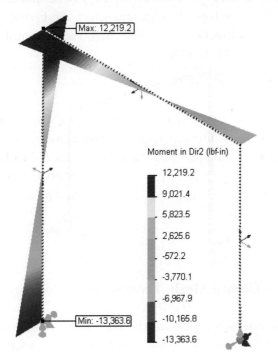

In SW Simulation, the local beam directions are set by the axial direction first, and then by the direction of the maximum (direction 2) and minimum principal directions of the cross-section area moment of inertia. For symmetrical cross-sections (like this example) the maximum and minimum inertias have the same value. Thus, the program arbitrarily picks one as direction 2. That means the choice of direction 1 or 2 may not be consistent for a single diagram display. Other FEA systems sometimes require the user to define the local directions for member output results. That is a confusing and error prone task that is avoided in SW.

This completes the consideration of the totally in-plane response of the stated settlement condition. Next the out of plane force will be superimposed to create a true space frame response.

Now, a normal joint force of 200 lb is superimposed on the previous case to see the out of plane displacements of this space frame:

(1) **External Loads → Force**. In the **Force panel → Select Joints**. Pick the top left joint and select the **Front plane** to set the reference direction. Pick the direction **Normal to**, and set the **Force** to 200 lb. **OK**.

(2) **Mesh → Create Mesh**. There are no mesh controls for frames. **Run**.

(3) The new reaction items are the base moments about the x-axis and the z-force at the bases. They are viewed with **Results →**

Define Displacement Plot $\rightarrow$ RMX: Reaction moment about x, etc.

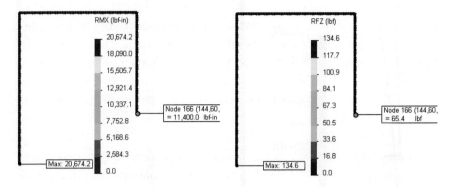

(4) **Results $\rightarrow$ Define Displacement Plot $\rightarrow$ URES: Resultant Displacement**, select **Advanced $\rightarrow$ Vector Plots**.

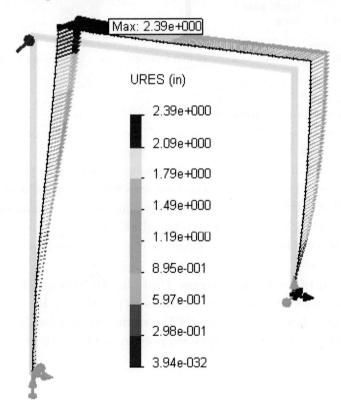

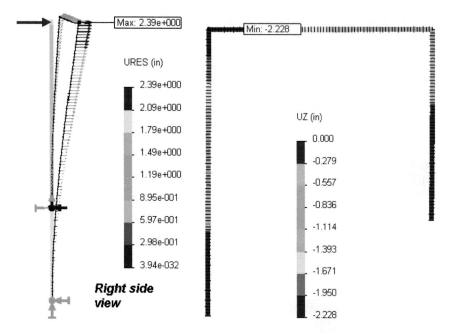

Fig. 10.12. Resultant and out of plane displacements for combined loadings.

The displacements have more than doubled with new displacement components normal to the plane of the frame. The out of plane displacement components are seen in Figure 10.12.

The out of plane stresses (UZ) are superimposed on the previous ones above. Therefore, the worst stresses will vary more around the perimeter of the cross-section. Some regions will have higher material stress levels while others have reduced values.

Re-check the worst case stress values for the combined case:

(1) **Results → Define Stress Plot.** In the **Stress Plot panel,** pull down **Worst case,** set **units** to psi. Click **OK.**

(2) Right click on the **Stress Plot name → Probe.** Select each node on the right leg from support to corner. Pick **graph icon,** click **OK.**

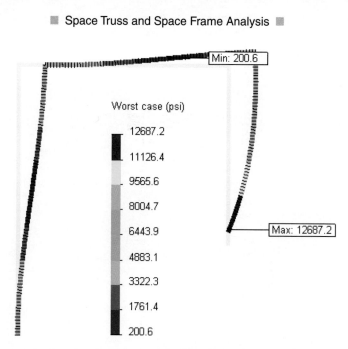

Other probed worst stress results for the combined support settlement and the out of plane normal force are given in Figure 10.13. The values have increased by about 50% and vary more with location.

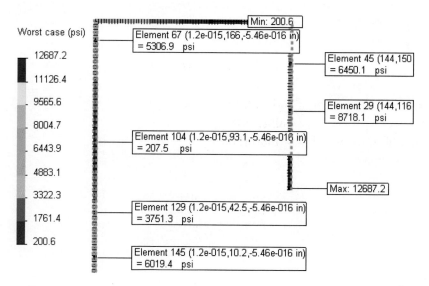

Fig. 10.13. Worst case element stress probe for the combined loadings.

11

Vibration Analysis

11.1. Introduction

A spring and a mass interact with one another to form a system that resonates at their characteristic natural frequency. If energy is applied to a spring-mass system, it will vibrate at its natural frequency. The level of a general vibration depends on the strength of the energy source as well as the damping inherent in the system. Consider the single degree of freedom system in Figure 11.1 that is usually introduced in a first course in physics or ordinary differential equations.

There, k is the spring constant, or stiffness, and m is the mass, and c is a viscous damper. If the system is subjected to a horizontal force, say $f(t)$, then Newton's law of motion leads to the differential equation of motion in terms of the displacement as a function of time, $x(t)$:

$$m\frac{d^2x}{dt^2} + c\frac{dx}{dt} + k\,x(t) = f(t),$$

which requires the initial conditions on the displacement, $x(0)$, and velocity, $v(0) = dx/dt(0)$. When there is no external force and no damping, then it is called free, undamped motion, or simple harmonic motion (SHM):

$$m\frac{d^2x}{dt^2} + k\,x(t) = 0.$$

217

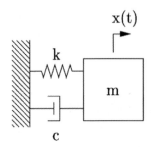

Fig. 11.1.   A spring-mass-damper single degree of freedom system.

The usual simple harmonic motion assumption is $x(t) = a\sin(\omega t)$ where a is the amplitude of motion and ω is the circular frequency of the motion. Then the motion is described by

$$[k - \omega^2 m]a\sin(\omega t) = 0, \quad \text{or } [k - \omega^2 m] = 0.$$

The above equation represents the simplest eigen-analysis problem. There you wish to solve for the eigenvalue, ω, and the eigenvector, a. Note that the amplitude, a, of the eigenvector is not known. It is common to scale the eigenvector to make the largest amplitude unity. The above scalar problem is easily solved for the circular frequency (eigenvalue),

$$\omega = 2\pi F_n = \sqrt{\frac{k}{m}},$$

which is related to the so called natural frequency, F_n, by $F_n = \omega/2\pi$.

From this, it is seen that if the stiffness increases, the natural frequency also increases, and if the mass increases, the natural frequency decreases. If the system has damping, which all physical systems do, its frequency of response is a little lower, and depends on the amount of damping. Numerous tabulated solutions for natural frequencies and mode shape can be found in Ref. [3]. They can be useful in validating finite element calculations.

Note that the above simplification neglected the mass of both the spring and the dampener. Any physical structure vibration can be modeled by springs (stiffnesses), masses, and dampers. In elementary models you use line springs and dampers, and point masses. It is typical to refer to such a system as a "lumped mass system". For a continuous part, both its stiffness and mass are associated with the same volume. In other words, a given volume is going to have a strain energy associated with its stiffness and a kinetic energy associated with its mass. A continuous part has mass and stiffness matrices that are of the same size (have the same number of DOF). The mass contributions therefore interact and cannot naturally be lumped to a single value at a point. There are numerical algorithms to accomplish such a lumped (or diagonal) mass matrix but it does not arise in the consistent finite element formulation.

11.2. Finite Element Vibration Studies

In finite element models, the continuous nature of the stiffness and mass leads to the use of square matrices for stiffness, mass, and damping. They can still contain special cases of line element springs and dampers, as well as point masses. Dampers dissipate energy, but springs and masses do not.

If you have a finite element system with many DOF then the above single DOF system generalizes to a displacement vector, $X(t)$ interacting with a square mass matrix, M, stiffness matrix, K, damping matrix C, and externally applied force vector, $F(t)$, but retains the same general form:

$$M\frac{d^2 X}{dt^2} + C\frac{dX}{dt} + KX(t) = F(t)$$

plus the initial conditions on the displacement, $X(0)$, and velocity, $v(0) = dX/dt(0)$. Integrating these equations in time gives a *time history solution*. The solution concepts are basically the same, they just have to be done using matrix algebra. The corresponding SHM, or free vibration mode ($C = 0, F = 0$) for a finite element system is

$$M\frac{d^2 X}{dt^2} + KX - X(t) = 0.$$

The SHM assumption generalizes to $X(t) = A\sin(\omega t)$ where the amplitude, A, is usually called the mode shape vector at circular frequency ω. This leads to the general matrix *eigenvalue problem* of a zero determinant:

$$|\boldsymbol{K} - \omega^2 \boldsymbol{M}| = 0.$$

There is a frequency, say ω_k, and mode shape vector, $\boldsymbol{A}_k$, for each degree of freedom, k. A matrix eigenvalue-eigenvector solution is much more computationally expensive that a matrix time history solution. Therefore most finite element systems usually solve for the first few natural frequencies. Depending on the available computer power, that may mean 10 to 100 frequencies. SW Simulation includes natural frequency and mode shape calculations as well as time history solutions.

Usually you are interested only in the first few natural frequencies. In SW Simulation, the default number of frequencies to be determined is five (that number is controlled via **Study → Properties → Options → Number of frequencies**). A zero natural (or slightly negative one) frequency corresponds to a rigid body motion. A part or assembly has at most six RBM of "vibration", depending on how or if it is supported. If a shell model is used the rotational DOF exist and the mass matrix is generalized to include the mass moments of inertia. For every natural frequency there is a corresponding vibration mode shape. Most mode shapes can generally be described as being an axial mode, torsional mode, bending mode, or general mode.

Like stress analysis models, probably the most challenging part of getting accurate finite element natural frequencies and mode shapes is to get the type and locations of the restraints correct. A crude mesh will give accurate frequency values, but not accurate stress values. The TK Solver case solver software contains equations for most known analytic solutions for the frequencies of mechanical systems. They can be quite useful in validating the finite element frequency results.

In Section 3.3, the stiffness matrix for a linear axial bar was given. It is repeated here along with its consistent mass matrix:

$$[k] = \frac{EA}{L} \begin{bmatrix} 1 & -1 \\ -1 & 1 \end{bmatrix}, \quad [m] = \frac{m}{6} \begin{bmatrix} 2 & 1 \\ 1 & 2 \end{bmatrix}, \quad m = \rho A L.$$

If you utilize a quadratic (three node) line element the corresponding element matrices are

$$[k] = \frac{EA}{3L} \begin{bmatrix} 7 & -8 & 1 \\ -8 & 16 & -8 \\ 1 & -8 & 7 \end{bmatrix}, \quad [m] = \frac{m}{30} \begin{bmatrix} 4 & 2 & -1 \\ 2 & 16 & 2 \\ -1 & 2 & 4 \end{bmatrix}.$$

11.3. Analytic Solutions for Frequencies

The analytic frequency and mode shape solutions for many parts with common geometries are found in a course on the vibration of continuous media. The geometries include axial bars, axial shafts in torsion, beams with transverse motion vibration, flat plates of various shapes, and thin shells of various shapes. Several examples of them are given in the "validation problems" set of examples presented along side the software tutorials.

Consider the longitudinal vibration of a bar. The results depend on which type of support is applied to each end of the bar. For one end restrained and the other end free the natural frequencies are

$$\omega_n = \frac{(2n-1)\pi c}{2L}, \quad c = \sqrt{\frac{E}{\rho}}, \quad n = 1, 2, 3, \ldots, \infty.$$

However, if both ends are restrained they are

$$\omega_n = \frac{n\pi c}{L}, \quad c = \sqrt{\frac{E}{\rho}}, \quad n = 1, 2, 3, \ldots, \infty.$$

This shows that for a continuous body there are, in theory, an infinite number of natural frequencies and mode shapes. Try a single quadratic element to model a fixed-fixed bar frequency. Restrain the two end DOF (the first and third row and column) of the above 3 by 3 matrices. Only a single DOF remains to approximate the first mode. Solve the restrained matrix eigen-problem: $[k] - \omega^2[m]| = 0$. The reduced terms in the matrices are

$$\frac{EA}{3L}[16] - \omega^2 \frac{\rho AL}{30}[16] = 0$$

so $\omega_1^2 = 10E/L^2\rho$ and $\omega_1 = \sqrt{10}c/L = 3.16c/L$ which is less than 1% error compared to the exact result. Adding more elements increases the accuracy of each frequency estimate, and also yields estimates of the frequencies associated with the additional DOF. For example, adding a second quadratic bar element gives a total of three un-restrained DOF. So you could solve for the first three frequencies. The value for ω_1 would be more accurate and you would have the first estimates of ω_2 and ω_3.

Usually, the masses farthest from the supports have the most effects on the natural frequency calculations. If you only care about the frequencies you could use split lines to build larger elements near the supports. For beams and shells, the transverse displacements are more important than the tangential rotational DOF.

11.4. Frequencies of a Curved Solid

To illustrate a typical natural frequency problem consider a brass 75 degree segment of an annulus solid having a thickness of 0.3 m, an average radius of 1.5 m, and a width of 1 m. The component is encastred (fixed) at one rectangular face. The thickness to width ratio is 0.3. That suggests that the study should be conducted with either a solid model or a thick shell model. Both types of elements will be used to indicate the range of uncertainty.

There is no simple analytic estimate to validate the study of a thick curved body. However, there is a simple cantilever beam frequency estimate that can give an estimate of the frequencies. The first frequency of such a thin beam is

$$\omega_1 = 1.732\sqrt{\frac{EI}{\rho AL^4}} = 1.732\sqrt{\frac{Eh^2}{12\rho L^4}}.$$

Here, the effect length, L, must be estimated. If you take the outer arc length as that length, the estimate is $\omega_1 = 48.4\,\text{Hz}$. Using the centerline gives 92.3 Hz.

Generally, the displacement degrees of freedom are more impor-tant in getting natural frequencies and mode shapes than are rotational DOF. Therefore, the solid study is probably best here. In vibration problems, the material located farthest from the supports are more important. You should use mesh control to create small elements in such regions. The modeling process is:

(1) Sketch and dimension the area. **Extrude** it to a thickness of 0.3 m.

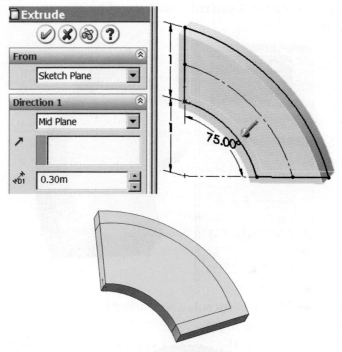

(2) Click on a curved face, **Insert Sketch**.
(3) Add a line and arc near the free edges farthest from the support, for later mesh control. **Insert → Curve → Split Line** (above).

Selecting the **SW Simulation Manager** icon: Right click on the top name to access **Study**, which opens the **Study panel**.

Assign a **Study name**, choose **Frequency** for the **Analysis type**. At this point **Solids** will appear in the Manager menu:

(1) Right click on it to apply material data. The component is to be made of brass.

(2) Pick **Apply Material to All** → **Material panel** → **From library files button** → **Copper Alloys** and select **brass**, set the **Units** to MKS.

Specify a finer mesh away from the support, and a crude mesh near the support:

(1) **Mesh** → **Mesh Control**, select small outer faces, set size to 0.06 m.

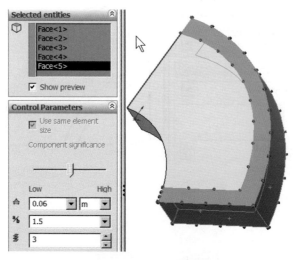

(2) **Mesh** → **Mesh Control**, select other faces, set size to 0.3 m.
(3) **Mesh** → **Create Mesh**.

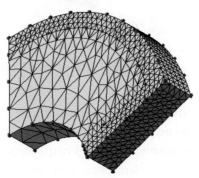

(4) Select **Fixtures** → **Immovable** and pick the support rectangles. **Run**.

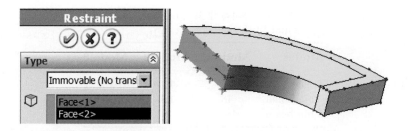

The **Run** → **Properties** were set to compute five modes and frequencies, but only the first three are summarized here. Select Results and display each mode in turn. Change views for better understanding as in Figure 11.2.

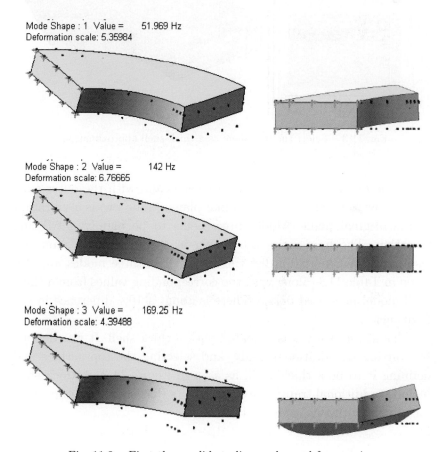

Fig. 11.2.   First three solid studies modes and frequencies.

Table 11.1. Natural frequencies (Hz) from solids and thick shells.

Model	Mode 1	Mode 2	Mode 3
Solid	52	142	169
Thick shell	46	126	155
Thin beam	48.4	—	—

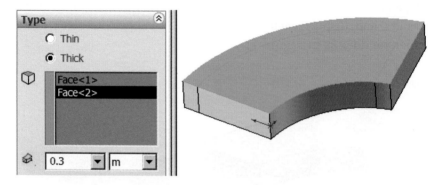

Fig. 11.3. Select the top plane as a thick shell approximation.

Mode one is like that of a cantilever beam, with the outer edge moving perpendicular to the original plane. Mode two is a vibration in the original plane. Mode three seems to be mainly a twisting vibration. The frequencies are shown in the figure text. You can also have SW Simulation list them. The first three modes are also given in Table 11.1, along with the corresponding values from a thick shell model presented below. There is about a 10% difference in the frequencies.

The above study was repeated with a thick shell. It was defined by starting an additional study, and selecting the top surface and defining it to be a thick shell, as seen in Figure 11.3. A thick shell includes additional transverse shear effects that should be important for a body with this aspect ratio.

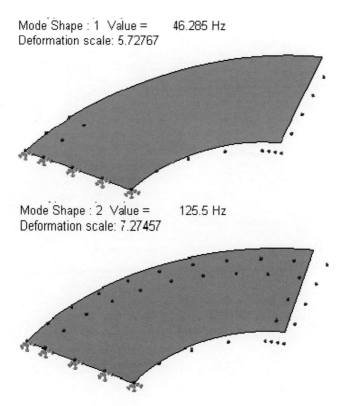

Mode Shape : 1 Value =   46.285 Hz
Deformation scale: 5.72767

Mode Shape : 2 Value =   125.5 Hz
Deformation scale: 7.27457

Fig. 11.4.   First two thick shell frequencies.

The same mesh controls were used to refine the model along its outer arc. Some results are in Figure 11.4 for the first two modes of vibration. Note that the thick shell approximation yields frequency estimates that are about lower that the full solid model.

Buckling Analysis

12.1. Introduction

There are two major categories leading to the sudden failure of a mechanical component: material failure and structural instability, which is often called buckling. For material failures you need to consider the yield stress for ductile materials and the ultimate stress for brittle materials.

Those material properties are determined by axial tension tests and axial compression tests of short columns of the material (see Figure 12.1). The geometry of such test specimens has been standardized. Thus, geometry is not specifically addressed in defining material properties, such as yield stress. Geometry enters the problem of determining material failure only indirectly as the stresses are calculated by analytic or numerical methods.

Predicting material failure may be accomplished using linear finite element analysis. That is, by solving a linear algebraic system for the unknown displacements, $K\delta = F$. The strains and corresponding stresses obtained from this analysis are compared to design stress (or strain) allowables everywhere within the component. If the finite element solution indicates regions where these allowables are exceeded, it is assumed that material failure has occurred.

The load at which buckling occurs depends on the stiffness of a component, not upon the strength of its materials. Buckling refers to the loss of stability of a component. The buckling mode is usually

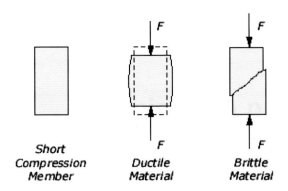

*Short
Compression
Member*   *F*
*Ductile
Material*   *F*
*Brittle
Material*

Fig. 12.1.   Short columns fail due to material failure.

independent of material strength. This loss of stability usually occurs within the elastic range of the material. The two phenomenon are governed by different differential equations [18]. Buckling failure is primarily characterized by a loss of structural stiffness and is not modeled by the usual linear finite element analysis, but by a finite element eigenvalue-eigenvector solution, $|K + \lambda_m K_F|\delta_m = 0$, where λ_m is the buckling load factor (BLF) for the m-th mode, K_F is the additional "geometric stiffness" due to the stresses caused by the loading, F, and δ_m is the associated buckling displacement shape for the m-th mode. The spatial distribution of the load is important, but its relative magnitude is not. The buckling calculation gives a multiplier that scales the magnitude of the load (up or down) to that required to cause buckling. The multiplier depends on the material modulus.

Slender or thin-walled components under compressive stress are susceptible to buckling. Most people have observed what is called "Euler buckling" where a long slender member subject to a compressive force moves lateral to the direction of that force, as illustrated in Figure 12.2. The force, F, necessary to cause such a buckling motion will vary by a factor of four depending only on how the two ends are restrained. Therefore, buckling studies are much more sensitive to the component restraints that in a normal stress analysis. The theoretical Euler solution will lead to infinite forces in very short columns, and that clearly exceeds the material ultimate

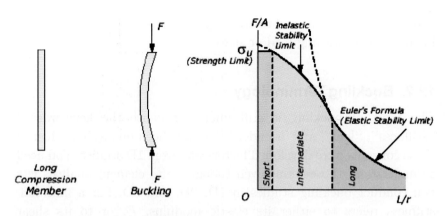

Fig. 12.2.   Long columns fail due to instability.

stress. In practice, Euler column buckling can only be applied to long columns and empirical transition equations are required for intermediate length columns. For very long columns the loss of stiffness occurs at stresses far below the material ultimate or yield stresses.

There are many analytic solutions for idealized components having elastic instability. About 75 of the most common cases are tabulated in the classic references [11, 15–17]. Euler long column buckling is quite sensitive to the end restraints. Figure 12.3 shows

Case	1	2	3	4	5
Constraints	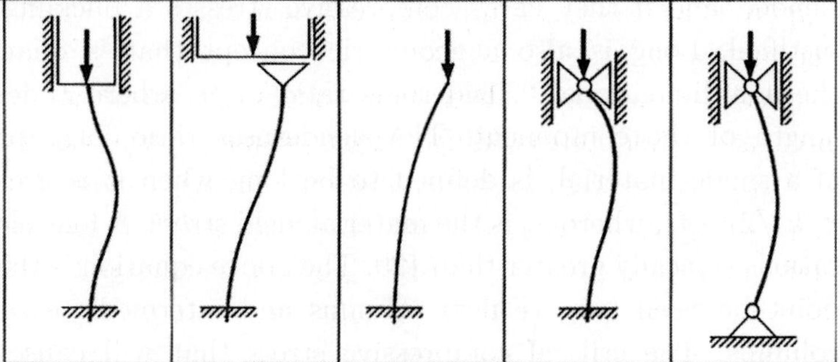				
k	4	1	.25	2.046	1

Fig. 12.3.   Restraints have a large influence on the critical buckling load.

five of several cases of end restraints and the associated k value used in computing buckling load or stress.

12.2. Buckling Terminology

The topic of buckling is still unclear because the keywords of "stiffness", "long" and "slender" have not been quantified. Most of those concepts were developed historically from 1D studies. You need to understand those terms even though finite element analysis lets you conduct buckling studies in 1D, 2D, and 3D. For a material, stiffness refers to either its elastic modulus, E, or to its shear modulus, $G = E/(2 + 2v)$ where v is Poisson's ratio.

Slender is a geometric concept addressing the ratio of a members length and a property of the cross-sectional area that is quantified by the radius of gyration. The radius of gyration, r, has the units of length and describes the way in which the area of a cross-section is distributed around its centroidal axis. If the area is concentrated far from the centroidal axis it will have a greater value of the radius of gyration and a greater resistance to buckling. A non-circular cross-section will have two values for its radius of gyration. The section tends to buckle around the axis with the smallest value. The radius of gyration, r, is defined as: $r = \sqrt{I/A}$, where I and A are the area moment of inertia, and area of the cross-section. For a circle of radius R, you obtain $r = R/2$. Solids can have regions that are slender, and if they carry compressive stresses a buckling study is justified. Long is also a geometric concept that is quantified by the non-dimensional "slenderness ratio" L/r, where L denotes the length of the component. The slenderness ratio, of a part made of a single material, is defined to be long when it is greater than $\pi/k\sqrt{2E/\sigma_y}$, where σ_y is the material yield stress. A long slenderness ratio is typically greater than 120. The above equation is the dividing point between long (Euler) columns and intermediate (empirical) columns. The critical compressive stress that will cause buckling always decreases as the slenderness ratio increases. The critical Euler buckling stress depends on the material, the slenderness ratio, and the end restraint conditions.

Table 12.1. Interpretation of the buckling load factor.

BLF	Buckling status	Remarks
> 1	Buckling not predicted	The applied loads are less than the estimated critical loads.
= 1	Buckling predicted	The applied loads are exactly equal to the critical loads. Buckling is expected.
< 1	Buckling predicted	The applied loads exceed the estimated critical loads. Buckling will occur.
$-1 < \text{BLF} < 0$	Buckling possible	Buckling is predicted if you reverse the load directions.
-1	Buckling possible	Buckling is expected if you reverse the load directions.
< -1	Buckling not predicted	The applied loads are less than the estimated critical loads, even if you reverse their directions.

12.3. Buckling Load Factor

The buckling load factor (BLF) is an indicator of the factor of safety against buckling or the ratio of the buckling loads to the currently applied loads. Table 12.1 illustrates the interpretation of possible BLF values returned by SW Simulation. Since buckling often leads to bad or even catastrophic results, you should utilize a high factor of safety for buckling loads (say BLF > 2).

12.4. General Buckling Concepts

Other 1D concepts that relate to stiffness are: axial stiffness, EA/L, flexural (bending) stiffness, EI/L, and torsional stiffness, GJ/L, where J is the polar moment of inertia of the cross-sectional area ($J = Iz = Ix + Iy$). Today, stiffness usually refers to the finite element stiffness matrix, which can include all of the above stiffness terms plus general solid or shell stiffness contributions. Analytic buckling studies identify additional classes of instability besides Euler buckling (see Figure 12.4). They include lateral buckling, torsional buckling, and other buckling modes. A finite element buckling study determines the lowest buckling factors and their

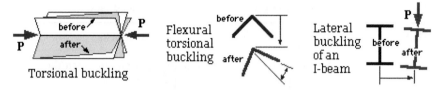

Fig. 12.4.   Some sample buckling mode shapes.

corresponding displacement modes. The amplitude of a buckling displacement mode, $|\delta_m|$, is arbitrary and not useful, but the shape of the mode can suggest whether lateral, torsional, or other behavior is governing the buckling response of a design.

12.5. Local Lateral Buckling of a Cantilever

Consider the plane stress analysis of a horizontal tapered cantilever beam subject to a transverse vertical load distributed over its free end face. The member was $L = 50$ inch long, $t = 2$ inch thick, and the depth, d, tapered from 3 inch at the load, to 9 inch at the fixed support. A study of the history of mechanics of materials shows that the concept of a *fixed*, or *cantilever*, or *encastre* support came from elementary beam theory. Originally it meant that a point at the neutral axis of a beam had both a zero displacement and rotation. That also implied that the support was capable of providing a reaction force vector and moment vector. There are several ways a support can provide a resisting force and moment to prevent a region of material from translating and rotating. Engineering practice has developed several standard symbols to represent a "fixed" support, rollers, pins, etc. Some of those symbols appear as icons in FEA restraints modules. You have to decide if those simplified support concepts are valid for your problem.

The use of finite element analysis in 2D or 3D requires you to always consider the effect of Poisson's ratio and how to avoid or model the point singularity in stresses that Poisson's ration causes due to restraint assumptions that you make. At a fixed wall in a 2D beam model the top fiber is typically being stretched horizontally. Due to Poisson's ratio, that same point wants to contract downward.

If that motion is prevented, as it is in the common assumption for a fixed support, then the vertical stress there must suddenly jump from basically zero to an extremely high value over a small area to develop the support force necessary to prevent that contraction. Then theoretical stress singularities develop in the 2D (and 3D) theory of elasticity solutions, but probably not in the physical entities. Since fixing the support edge will cause false infinite stresses at the two corners, a more realistic support condition was used in this example by including a semi-circular segment of the plate to which the beam is welded. The edge of that curved plate segment was fixed instead. That removes a singularity due to Poisson's ratio. However, a weaker singularity remains because a sharp re-entrant corner was used. A simple fillet would remove that singularity.

The plane stress model was built, loaded, restrained as described above and solved. The deformed shape in Figure 12.5 is what you would expect and predict from simple beam theory. The predicted flexural stresses are almost constant along the length instead of increasing as they approach the support. That is because the beam is tapered. The moment of inertia of the beam increases with the cube of the depth. Thus, it grows faster than the bending moment due to the end load.

Even though the stress distribution looks different from a constant thickness beam, the stresses are quite low here. Both the von Mises and the maximum shear stress material failure criterion

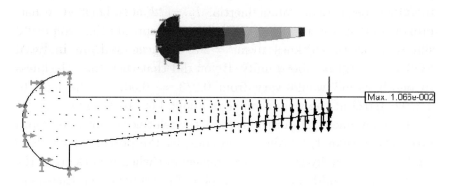

Fig. 12.5.   Plane stress deflection magnitudes and directions.

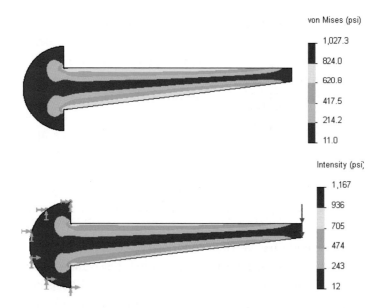

Fig. 12.6.   von Mises (top) and maximum shear stress failure criteria.

(Figure 12.6) are around 1 ksi. Comparing either one to the material yield point of about 90 ksi you can see by inspection that the material factor of safety is about 90 to 100, far above the minimum required value of unity.

This very high material factor of safety probably suggests (incorrectly) that a simple redesign will save material, and thus money. The load carrying capacity of a beam is directly proportional to its geometric moment of inertia, $I_z = td^3/12$. Thus, it also is proportional to its thickness, t. Therefore, it appears that you could simply reduce the thickness from $t = 2$ to 0.2 inches and your material FOS would still be above unity. If you did that then the "thickness to depth ratio" would vary from $0.2/3 = 0.067$ at the load to $0.2/9 = 0.022$ at the wall, a range of about $1/15$ to $1/45$.

If a component has a region where the relative thickness to depth ratio of less than $1/10$ you should consider the possibility of "*local buckling*". It usually is a rare occurrence, but when it does occur the results can be sudden and catastrophic. To double check the safety of reducing the thickness you should add a second study that utilizes the SW Simulation buckling feature to determine the lowest buckling load. To do that:

(1) Right click on the **Simulation → New Study** to open the **Study panel**. Assign a new **Study name**, select Buckling as the **Type of analysis**.
(2) To use the same loads and restraints from the prior static study **drag** the **External Loads** from the first study and **drop** them into the second one (except for any normal to the plane). Likewise, **drag and drop** the first shell **Materials** into the second study.
(3) Create a new finer mesh. Right click on the **Part name → Run**.

A buckling, or stability, analysis is an eigen-problem. The magnitude of the scalar eigen-value is called the *"buckling load factor"*, BLF. The computed displacement eigen-vector is referred to as the *"buckling mode"* or mode shape. They are only relative displacements. Usually they are presented in a non-dimensional fashion where the displacements range from zero to ±1. In other words, the actual value or units of a buckling mode shape are not important. Still, it is wise to carry out a visual check of the first buckling mode:

(1) When the solution completes, pick **Displacements → Plot 1** and examine the resultant relative displacement URES. Note that the displacement contour curves in Figure 12.7 are inclined to the long axis of the beam instead of being vertical as before in Figure 12.5.
(2) Use **Settings** to get a plot of the displacement along with the undeformed shape, and rotate the display to an out-of-plane view, as in Figure 12.8.

The displacement magnitudes shown in the figures are simply relative. Most FEA systems scale the buckling mode shapes to range from zero to ±1. The signs mean that free end moved out of plane in the positive z-direction, while the bottom corner, near the support, moved in the opposite direction. From Figure 12.8 you see that under the vertical load the (very thin) beam buckled mainly sideways (perpendicular to the load) rather than downward. This is

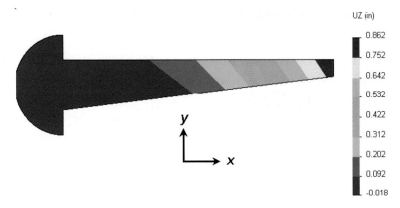

Fig. 12.7. Buckling mode displacement values (normal to surface).

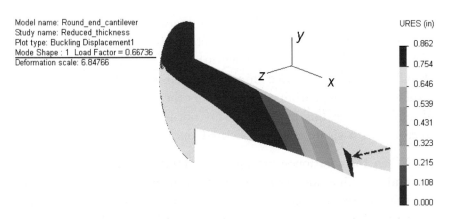

Fig. 12.8. Relative lateral (out of plane) buckling mode displacements.

an example of lateral buckling. That is typical of what can happen to very thin regions.

Next, the question is: how large must the end load be to cause such motion, and failure? To see the magnitude of the BLF (eigenvalue): right click on **Results** → **List Buckling Mode Factors**. In the **List Modes panel**, Figure 12.9, read the BLF value of about 0.67. The buckling load factor is an indicator of the factor of safety against buckling or the ratio of the buckling loads to the currently applied loads. Since buckling often leads to bad or even catastrophic results, you should utilize a high buckling factor of

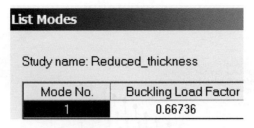

Fig. 12.9. First buckling mode load factor.

safety (at least > 2) for buckling loads. Instead, the study shows that only about 2/3 of the planned load will cause this member to fail by lateral buckling due to loss of stiffness in the out of plane direction.

To understand why this failure has occurred, you must reconsider the thickness reduction or plan to add lateral bracing members and repeat the buckling study. Remember that the geometric moment of inertia of the beam about the vertical (y) axis is $I_y = dt^3/12$. It is a measure of the lateral bending (and buckling) resistance. By reducing the thickness, t, by a factor of 10 the original I_z (and the in-plane bending resistance) went down by the same factor of 10, but I_y (and the out-of-plane bending resistance) went down by a factor of 1,000.

To illustrate the importance of lateral bracing, consider two lateral support options: (1) only the vertical free tip edge is restrained against motion normal to the beam; and (2) the beam is fully supported laterally over its full tapered face area. The new BLF, listed in Figure 12.10 increase drastically.

Figure 12.11 shows that the lateral motion in the first case occurs mainly at the mid-span. The BLF is then about 1.6 which means that the applied load would have to be increased by that factor for

Mode No.	Buckling Load Factor
1	-1.5621

Mode No.	Buckling Load Factor
1	-1216.9

Fig. 12.10. BLF from lateral support of the tip (left) and full beam face.

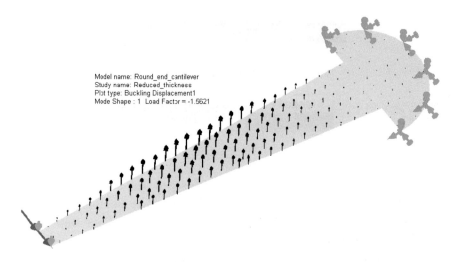

Model name: Round_end_cantilever
Study name: Reduced_thickness
Plot type: Buckling Displacement1
Mode Shape : 1 Load Factor = -1.5621

Fig. 12.11.   Buckling mode with lateral support at the tip edge.

buckling to theoretically occur. In reality, that value was computed with a perfectly flat beam. If a small lateral miss-alignment was included in the construction of the beam (as happens in the real world), the BLF would be smaller. If the beam is fully supported laterally, as in the second case, then the listed BLF is very high (Figure 12.10) because then the displacements must be in the plane of the beam instead of transverse to it.

13

Concepts of Thermal Analysis

13.1. Introduction

There are three different types of **heat transfer**: conduction, convection, and radiation. A temperature difference must exist for heat transfer to occur. Heat is always transferred in the *direction* of decreasing temperature. Temperature is a scalar, but heat flux is a vector quantity. The thermal variables and boundary conditions relate to the displacements and stress in an axial bar through the analogy as summarized in Table 13.1.

Conduction takes place within the boundaries of a body by the diffusion of its internal energy. The temperature within the body, T, is given in units of degrees Celsius [C], Fahrenheit [F], Kelvin [K], or Rankin [R]. Its variation in space defines the temperature gradient vector, ∇T, with units of [K/m] say. The heat flux vector, q, per unit area is define by Fourier's Conduction Law, as the thermal conductivity matrix, k, times the negative of the temperature gradient, $q = -k\nabla T$. The integral of the heat flux over an area yields the total heat flow for that area.

Thermal conductivity has the units of [W/m-K] while the heat flux has units of [W/m^2]. The conductivity, k, is usually only known to three or four significant figures. For solids it ranges from about 417 W/m-K for silver down to 0.76 W/m-K for glass. A perfect insulator material ($k \equiv 0$) will not conduct heat; therefore the heat flux vector must be parallel to the insulator surface. A plane of symmetry (where the geometry, k values, and heat sources are mirror

Table 13.1.   Terms of the 1D thermal-structural analogy.

Thermal analysis item, [units], symbol	Structural analysis item, [units], symbol
Unknown: Temperature [K], T	Unknown: Displacements [m], u
Gradient: Temperature gradient [K/m], ∇T	Gradient: Strains [m/m], ε
Flux: Heat flux [W/m^2], q	Flux: Stresses [N/m^2], σ
Source: Heat source for point, line, surface, volume [W], [W/m], [W/m^2], [W/m^3], Q	Source: Force for point, line, surface, volume [N], [N/m], [N/m^2], [N/m^3], g
Indirect restraint: Convection	Indirect restraint: Elastic support
Restraint: Prescribed temperature [K], T	Restraint: Prescribed displacement [m], u
Reaction: Heat flow resultant [W], H	Reaction: Force component [N], F
Material Property: Thermal conductivity [W/m-K], k	Material Property: Elastic modulus [N/m^2], E
Material Law: Fourier's law	Material Law: Hooke's law

images) acts as a perfect insulator. In finite element analysis, all surfaces default to perfect insulators unless you give a specified temperature, a known heat influx, a convection condition, or a radiation condition.

Convection occurs in a fluid by mixing. Here we will consider only *free convection* from the surface of a body to the surrounding fluid. *Forced convection*, which requires a coupled mass transfer, will not be considered. The magnitude of the heat flux normal to a solid surface by free convection is $q_n = hA_h(T_h - T_f)$ where h is the convection coefficient, A_h is the surface area contacting the fluid, T_h is the convecting surface temperature, and T_f is the surrounding fluid temperature, respectively. The units of h are [W/m^2-K]. Its value varies widely and is usually known only from one to four significant figures. Typical values for convection to air and water are $5-25$ and $500-1000$ W/m^2-K, respectively.

Radiation heat transfer occurs by electromagnetic radiation between the surfaces of a body and the surrounding medium. It is a highly nonlinear function of the absolute temperatures of the body and medium. The magnitude of the heat flux normal to a solid surface by radiation is $q_r = \varepsilon \sigma A_r (T_r^4 - T_m^4)$. Here T_r is the absolute temperature of the body surface, T_m is the absolute temperature of the surrounding medium, A_r is the body surface area subjected to radiation, $\sigma = 5.67 \times 10^8 \, \text{W/m}^2 - K^4$ is the Stefan–Boltzmann constant, and ε is a surface factor ($\varepsilon = 1$ for a perfect black body).

Transient, or unsteady, heat transfer in time also requires the material properties of specific heat at constant pressure, c_p in [kJ/kg-K], and the mass density, ρ in [kg/m^3]. The specific heat is typically known to 2 or 3 significant figures, while the mass density is probably the most accurately known material property with 4 to 5 significant figures.

The one-dimensional governing differential equation for transient heat transfer through an area A, of conductivity k_x, density ρ, specific heat c_p with a volumetric rate of heat generation, Q, for the temperature T at time t is $\partial(k_x \partial T/\partial x)/\partial x + Q(x) = \rho c_p \partial T/\partial t$, for $0 \leq x \leq L$ and time $t \geq 0$. It requires initial conditions to describe the beginning state, and boundary conditions for later times. For a steady state condition ($\partial T/\partial t = 0$) the typical boundary conditions of one of the following:

(1) T prescribed at 0 and L, or
(2) T prescribed at one end and a heat source at the other, or
(3) T prescribed at one end and a convection condition at the other, or
(4) A convection condition at one end and a heat source at the other, or
(5) A convection condition at both ends.

In the 3D case the differential equation becomes the anisotropic Poisson Equation (see Chapter 15). That is, the above diffusion term (second derivatives in space) is expanded to include derivatives with respect to y and z, times their corresponding thermal conductivity values.

Table 13.2.   Isotropic thermal properties.

Symbol	Label	Item	Application
ρ	DENS	Mass density	Transient
c	C	Specific heat, at constant pressure	Transient
k	KX	Thermal conductivity	Steady state, transient

Table 13.3.   Anisotropic thermal properties in principal material directions.

Symbol	Label	Item
ρ	DENS	Mass density
c	C	Specific heat, at constant pressure
k_x	KX	Thermal conductivity in material X direction
k_y	KY	Thermal conductivity in material Y direction
k_z	KZ	Thermal conductivity in material Z direction

13.2. Thermal Analysis Input Properties

Tables 13.2 and 13.3 list the isotropic and orthotropic thermal material properties available in SW Simulation, respectively. For anisotropic materials you usually need to utilize the custom material definition process (page 94), and the material direction definitions given below.

Only the conductivities are theoretically needed for a steady state study, but SW Simulation always requests the mass density. Any transient (time dependent) thermal analysis involves the product of the mass density and specific heat, as seen in the above equation.

13.3. Finite Element Thermal Analysis

The finite element method creates a set of algebraic equations by using an equivalent governing integral form that is integrated over a mesh that approximates the volume and surface of the body of interest. The mesh consists of elements connected to nodes. In a thermal analysis, there will be one simultaneous equation for each node. The unknown at each node is the temperature. Today, a typical thermal mesh involves 20,000 to 100,000 nodes and thus temperature equations. The restraints are specified temperatures (or a convection condition since it includes a specified fluid temperature). The reactions are the resultant heat flow that is necessary to maintain a specified temperature. All other conditions add load or source terms. The default surface condition is an insulated boundary, which results in a zero source (load) term. The assembled matrix equations for thermal equilibrium have exactly the same partitioned form as the structural systems of Section 2.5:

$$
\begin{bmatrix} K_{uu} & K_{ug} \\ K_{gu} & K_{gg} \end{bmatrix} \begin{Bmatrix} T_u \\ T_g \end{Bmatrix} = \begin{Bmatrix} F_g \\ F_u \end{Bmatrix},
$$

where now T_g represents the given (restrained) nodal temperatures, F_g represents the known resultant nodal heat power (heat flow) at the node. This system of equations is solved for unknowns T_u just as described in Section 2.5. The thermal reactions, F_u, at the given temperature nodes represent the total heat flow, in or out, necessary to maintain the given temperatures, T_g. From the above structural-thermal analogy, the matrix equations of a linear (temperature interpolation) conducting element (from Sections 2.3 and 2.4) is

$$
k \begin{bmatrix} 1 & -1 \\ -1 & 1 \end{bmatrix} \begin{Bmatrix} T_1 \\ T_2 \end{Bmatrix} = \begin{Bmatrix} F_1 \\ F_2 \end{Bmatrix},
$$

where $k \equiv k_x A/L$ may be referred to as the thermal stiffness of the rod of length, L, area, A, and thermal conductivity k_x. In this case, T corresponds to a nodal temperature, and F corresponds to the resultant nodal heat power from the various heat sources. The thermal load (source) items for steady state analysis are given in Table 13.4. Both convection and radiation require inputs of the estimated surface conditions.

Table 13.5 gives typical convection coefficients values. Note that there is a wide range in such data. Therefore, you will often find it necessary to run more than one study to determine the range of answers that can be developed in your thermal study.

Table 13.4.　Loads for steady state thermal analysis.

Load type	Geometry	Required input
Convection	Faces	Film coefficient and bulk temperature
Heat Flux	Faces	Heat flux (heat power/unit area) value
Heat Power	Pts, edges, faces, parts	Total heat power value (rate of heat generation per unit volume times the part volume)
Insulated (Adiabatic)	Faces	None. This is the *default condition* for any face not subject to one of the three above conditions
Radiation	Faces	Surrounding temperature, emissivity values, and view factor for surface to ambient radiation

Table 13.5.　Typical heat convection coefficient values, h, $[\text{W/m}^2 \text{ K}]$.

Fluid medium	h
Air (natural convection)	5–25
Air/superheated steam (forced convection)	10–500
Oil (forced convection)	60–1,800
Steam (condensing)	5,000–120,000
Water (boiling)	2,500–60,000
Water (forced convection)	300–6,000

Having supplied all the restraints, loads, and properties you can run a thermal analysis and continue on to post-processing and documenting the results.

Table 13.6 gives the thermal restraints items for steady state analysis. Most programs offer only a temperature restraint. SW Simulation also offers the ability to define a non-ideal material interface, as in Figure 13.1. That is often needed in practice and is

Table 13.6.　Restraints in steady state thermal analysis.

Restraint	Geometric entities	Required input
Temperature	Vertexes, edges, faces and parts	Temperature value
Contact resistance	Two contacting faces. See discussion	Total or unit thermal resistance

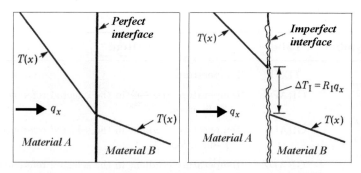

Fig. 13.1.   Ideal and thermal contact resistance interfaces.

Table 13.7.   Typical contact resistance values, $R \times e4$, $[\text{m}^2\,\text{K/W}]$.

Contact pressure	Moderate	$100\,\text{kN/m}^2$	$1e4\,\text{kN/m}^2$
Aluminum/aluminum/air	0.5	1.5–5.0	0.2–0.4
Copper/copper/air	0.1	1–10	0.1–0.5
Magnesium/magnesium/air	0.5	1.5–3.5	0.2–0.4
Stainless steel/stainless steel/air	3	6–25	0.7–4.0

referred to as a contact resistance. It basically defines a temperature jump across an interface for a given heat flux through the interface. The necessary resistance input, R, depends on various factors. Table 13.7 gives typical R values, while Table 13.8 cites values of its reciprocal, the conductance.

The temperature often depends only on geometry. The heat flux, and the conthermal reaction, always depends on the material thermal conductivity. Therefore, it is always necessary to examine both the

Table 13.8.   Typical contact conductance values, C, $[\text{W/m}^2\text{K}]$.

Contacting faces (pressure unknown)	Conductance
Aluminum/aluminum/air	2,200–12,000
Ceramic/ceramic/air	500–3,000
Copper/copper/air	10,000–25,000
Iron/aluminum/air	45,000
Stainless steel/stainless steel/air	2,000–3,700
Stainless steel/stainless steel/vacuum	200–1,100

Table 13.9.   Thermal analysis output options.

Symbol	Label	Item
T	TEMP	Temperature
$\dfrac{\partial T}{\partial x}$	GRADX	Temperature gradient in the selected reference X-direction
$\dfrac{\partial T}{\partial y}$	GRADY	Temperature gradient in the selected reference Y-direction
$\dfrac{\partial T}{\partial z}$	GRADZ	Temperature gradient in the selected reference Z-direction
$\lvert \nabla T \rvert$	GRADN	Resultant temperature gradient magnitude
q_x	HFLUXX	Heat flux in the X-direction of the selected reference geometry
q_y	HFLUXY	Heat flux in the X-direction of the selected reference geometry
q_z	HFLUXZ	Heat flux in the X-direction of the selected reference geometry
q	HFLUXN	Resultant heat flux magnitude

temperatures and heat flux to assure a correct solution. The heat flux is determined by the gradient (derivative) of the approximated temperatures. Therefore, it is less accurate than the temperatures. The user must make the mesh finer in regions where the heat flux vector is expected to rapidly change its value or direction. The heat flux should be plotted both as magnitude contours, and as vectors. The items available for output after a thermal analysis run are given in Table 13.9.

The temperatures should be plotted as discrete color bands or as contour lines. The temperature contours should be perpendicular to insulated boundaries. Near surfaces with specified temperatures, the contours should be nearly parallel to the surfaces. These "eyeball" checks are illustrated in Figure 13.2.

The heat flux vectors should be parallel to insulated surfaces. They should be nearly perpendicular to surfaces with a specified constant temperature. Those flux checks are illustrated in Figure 13.3. These remarks on insulated boundaries do not apply if the material is anisotropic with the principal material directions inclined relative to the insulated surface (as will be seen later).

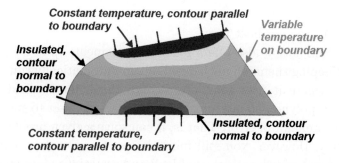

Fig. 13.2. Guidelines for checking temperatures in isotropic materials.

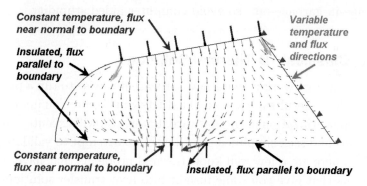

Fig. 13.3. Graphical checks for heat flux in isotropic materials.

The exact temperature gradient is discontinuous at an interface between different materials because their thermal conductivities will be different. Pretty continuous color contours (the default) tend to prevent these important engineering checks. The temperature and temperature gradient vector can depend only on the geometry in some problems. Written results should not be given with more significant figures than the material input data. For heat transfer problems that is typically three or four significant figures.

In SW Simulation it is possible to list, sum, average, and graph results along selected edges, lines, curves or surfaces. Thus, you should plan ahead and add "split lines" to the mesh where you expect to find such graphs informative. The thermal reaction heat flows is available in SW Simulation while viewing the heat flux result plots. A thermal reaction is obtained via **Results → Define Thermal**

Plot → **Heat flux resultant** → **List selected** where the sum entry gives the net heat flow.

SW Simulation also offers p-adaptive elements (p is for polynomial). Keeping the mesh unchanged, it can automatically run a series of cases where it uses complete second, third, fourth, and finally fifth order polynomial interpolations. It allows the user to specify the allowable amount of error. That is, it can solve a given problem quite accurately. However, you still must define the geometry, materials, load and restraint locations, and load and restraint values as well as interpret the results properly. You still have the age old problem of garbage-in garbage-out, so avoid computer aided stupidity.

13.4. Classical 1D Thermal Solutions

When you start working with new software it is wise to run a problem for which the results are known. That lets you be sure you understand the proper utilization of the software. Later, you will execute complex problems and seek simplified solutions in an attempt to validate your study. There are a few well know thermal problems that have simple solutions that give you some insight into heat transfer solutions and are easily verified with a SW Simulation analysis. The first of these is a planar wall with a temperature difference on each side.

This is often approximated as a semi-infinite wall, having a unit area $A = 1$ and thickness of L, which reduces the mathematical problem to a one-dimensional study. The solution [5] shows that the temperature through the wall is linear in space: $T(x) = T_{in}(1 - x/L) + T_{out}x/L$. Therefore, the heat flux, per unit area, is constant: $q = -kA\,dT/dx = k(T_{in} - T_{out})/L$. Note that the temperature distribution depends only on the shape, but the heat flux always depends on the material. Any finite element model will give the exact result [2].

The heat transfer through a wall will be illustrated by a SW Simulation model. It could be solved with a single layer of elements through the wall. Here it is assumed that the analytic solution is not known, so a large number of unknowns are used to clearly illustrate the response. The alloy steel ($k = 6.69e\text{-}4\,\text{BTU/in-s-F}$) wall is five

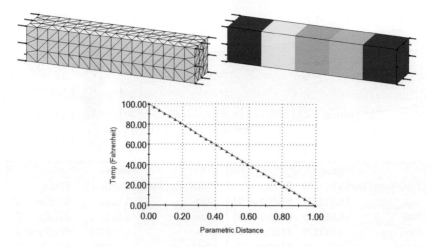

Fig. 13.4.   Linear temperatures of a homogeneous wall.

inches thick. A unit cross-sectional area is used. The inner (left) side
is kept at 100 F while the outer side is at 0 F. Those two restraints
must be explicitly applied. The other four faces of the body are
planes of symmetry and are automatically treated as insulated. The
mesh is shown along with the resulting linear temperature drop
distribution. The linear temperature change with position is clearly
seen in Figure 13.4.

Note that at a position 40% through the wall the temperature
difference has dropped 40% to 60 F. This result will be compared to
a cylindrical wall later. The heat flux should be constant. Constant
values do not contour well so the contour bounds must be set to
give a reasonable plot. The flux values at the inlet and outlet faces
are selected and listed in tables shown in Figure 13.5. It shows that
each square inch of the outer wall requires about 0.0134 BTU/s of
power to maintain the outer temperature. For a planar wall made up
of constant thickness layers of different materials the heat flux must
still remain constant, but the temperature difference will occur as
linear changes from one interface to the next. The linear distribution
of temperature is more easily seen with a graph along one edge of
the mesh.

Another well known heat transfer problem with a simple analytic
solution is that of radial conduction through an infinite pipe, or

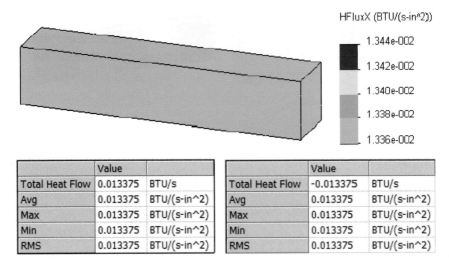

	Value	
Total Heat Flow	0.013375	BTU/s
Avg	0.013375	BTU/(s-in^2)
Max	0.013375	BTU/(s-in^2)
Min	0.013375	BTU/(s-in^2)
RMS	0.013375	BTU/(s-in^2)

	Value	
Total Heat Flow	-0.013375	BTU/s
Avg	0.013375	BTU/(s-in^2)
Max	0.013375	BTU/(s-in^2)
Min	0.013375	BTU/(s-in^2)
RMS	0.013375	BTU/(s-in^2)

Fig. 13.5. Constant heat flux gives equal and opposite heat flows.

curved wall. In that case, the temperature difference varies in a logarithmic manner through the wall thickness: $T(r) = T_{in} + (T_{out} - T_{in}) \ln(r/r_{in}) / \ln(r_{out}/r_{in})$. That means that the heat flux per unit area must decrease through the wall, since it passes through more material: $q(r) = k(T_{in} - T_{out})/(r \ln(r_{out}/r_{in}))$. The example here [4], will be for an alloy steel ($k = 6.69e\text{-}4$ BTU/in-s-F) pipe with an inner radius of 4 inches and with a thickness of 5 inches. Thus, it is very similar to the previous example having inner and outer temperatures of 100 F and 0 F, respectively. In this case, each of those restraints is applied to cylindrical faces. The other four faces of the solid radial wedge are insulated and do not require specific action. A fine mesh, the resulting temperature contours, and the radial variation of the temperature are given in Figure 13.6. The contour plot there might appear to again be linear, but the graph of the temperature along a radial edge is actually logarithmic. Compared to Figure 13.4, you see that at a distance of 40% through the wall the temperature has dropped more than 40% to about 50 F.

The non-constant nature of the corresponding heat flux is seen in the contour plot and in the radial edge heat flux graph of Figure 13.7. The exact inner and outer heat flux values are 0.0206

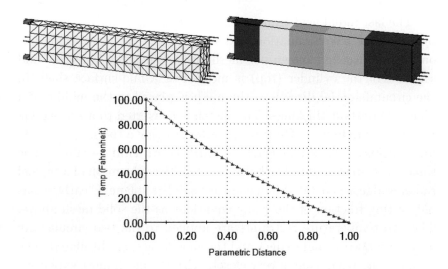

Fig. 13.6.   Logarithmic radial temperature through a cylindrical wall.

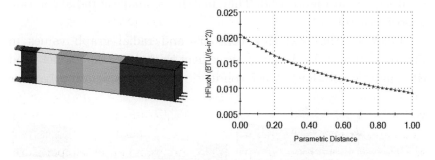

	Value	
Total Heat Flow	0.0071914	BTU/s
Avg	0.02061	BTU/(s-in^2)
Max	0.020618	BTU/(s-in^2)
Min	0.020602	BTU/(s-in^2)
RMS	0.02061	BTU/(s-in^2)

	Value	
Total Heat Flow	-0.0071957	BTU/s
Avg	0.0091622	BTU/(s-in^2)
Max	0.0091631	BTU/(s-in^2)
Min	0.0091616	BTU/(s-in^2)
RMS	0.0091622	BTU/(s-in^2)

Fig. 13.7.   Radial heat flux in a cylindrical wall, and the heat flow.

and 0.00917 BTU/s-in^2, respectively. Those heat flux values agree very well with the graph end points. The total radial heat flow, qA, through the cylinder is constant since the conducting area increases as the heat flow per unit area, q, decreases.

The last radial heat transfer example could have also been solved by using the SW Simulation mid-surface shell element, which has one temperature unknown per mesh node. When the 5 degree solid segment of the cylinder (top) is meshed as a mid-surface shell (in the circumferential direction) the mesh is placed in the middle of a plane of constant thickness. The sketch is converted to a surface via **Insert → Surface → Planar** and after starting a thermal study its unit thickness was set in **Part → Edit Definition → Thin**. The same allow steel material as before is used. Here the mesh of a typical radial wedge is generated in a constant axial (z) plane. Clearly, it has only a tiny fraction of the many equations as the solid mesh above. The two temperature restraints are applied to the two circular arc edges. The two straight edges and the shell face(s) are insulated. The temperature results agree very closely with the much more expensive solid computations. That is easily seen by examining the temperature results given in Figure 13.8. They match the analytic radial solution almost exactly.

Likewise, the heat flux contours and radial graph values in Figure 13.9 are also in close agreement with the solid model (and the analytic solution). Most conduction problems also involve free

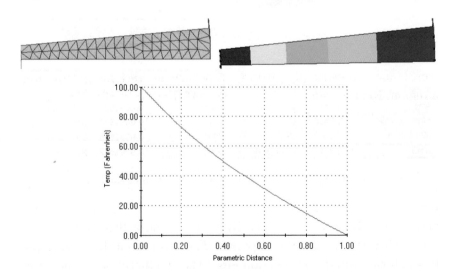

Fig. 13.8.   Pipe segment temperatures from mid-surface shell mesh.

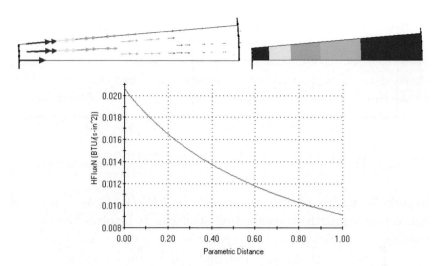

Fig. 13.9.   Mid-surface shell heat flux result for the pipe.

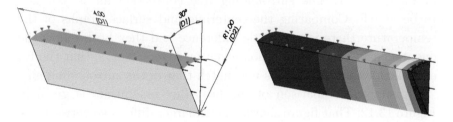

Fig. 13.10.   Circular rod with an end temperature and convection.

convection. That usually gives a steeper change in temperature over a region. Next a segment of a circular rod (Figure 13.10) is examined where the length is only two times the diameter.

That is near the lower limit where you might want to expect a one-dimensional approximation to be accurate. Convection occurs on the outer surface while one end is kept at 100 F. The other three symmetry surfaces in the model are insulated. Any wedge angle could have been used, but a value of 30 degrees was picked to give good element aspect ratios.

Myers [8] gives the one-dimensional solution for a rod conducting heat along its interior and convection that heat away at its surface. The temperature is shown to change with axial position, x, as a

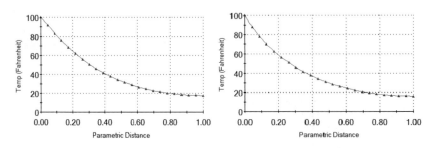

Fig. 13.11.   Center (left) and surface temperature, for a small m value.

hyperbolic cosine of mx, where $m^2 = hPL/kAL$, is a ratio of convection strength to conduction strength. It involves the surface convection coefficient, h, the perimeter, P, of the conducting area, A, over the length L, and the material thermal conductivity, k. Typical temperature distributions, for a low value of m are seen in Figure 13.11. The surrounding free convection air is assumed to be at $0\,\mathrm{F}$. Comparing the centerline and surface graphs of the temperature there is very little difference and they both follow the one-dimensional approximation given by Myers. Notice that the far end plane temperature does not match that of the surrounding air.

A similar comparison of the heat flux magnitude is given in Figure 13.12. That figure shows a much larger difference between the

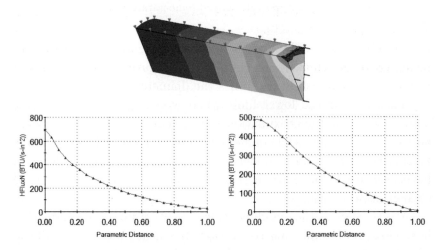

Fig. 13.12.   Surface and center heat flux magnitude results, for a small m.

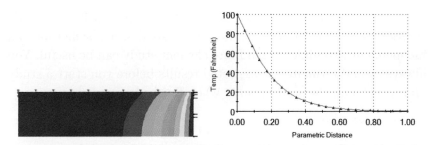

Fig. 13.13.   Center temperature for convection, h, increased by ten.

centerline and surface heat flux. But the average of the two graphs is still quite close to the analytic approximation given by Myers.

It is not uncommon for the user supplied convection coefficient to be in error due to measurement errors or errors occurring in a unit's conversion. As an example, the above study was re-run with the convection coefficient increased by a factor of 10. The convection heat transfer mode was increased relative to the conduction mode (m was approximately tripled). The new temperatures, in Figure 13.13, are significantly different from those of Figure 13.11.

The surface and centerline temperature graphs are still about the same and still follow the hyperbolic cosine change given by Myers. However, the temperatures in the distal half of the bar have dropped to rapidly approach, or match the temperature of the surrounding air.

The convection coefficient has lower and upper bounds, of 0 and ∞. The two bounding values have different physical effects in a study. A low value of h causes the surface to approach an insulated state, while a high value causes the surface to approach a restraint of a specified temperature. The latter state is what is seen in Figure 13.13. The distal end of the part is responding as if it had a restraint temperature of $0\,\mathrm{F}$ applied to it. These two limits on h are also reflected in terms of the temperature contour lines. The lower limit causes the contour lines to approach being perpendicular to the surface as they do for insulated boundaries. Likewise, the upper limit causes the temperature contours to approach being parallel to the surface as they would if it was subjected to a constant temperature restraint. These examples illustrate the value of using analytic approximations to estimate and validate the results from a finite element study. The first example also shows that if an analytic solution is not available for validating a solid study sometimes an independent two-dimensional finite element study can be useful. You always should estimate the expected results before you start a study and to validate the study results when finished.

13.5. Heat Transfer with an Orthotropic Material

It is becoming more common to encounter materials which have properties that are directionally dependent (anisotropic). A common case is that of orthotropic materials that have their properties completely defined in terms of three perpendicular directions. Those three principal material directions are usually defined by a user defined coordinate system. The input reference system provides the data necessary to compute the direction cosines between the material directions and the global x–y–z-axes. That defines a coordinate transformation matrix, say T, that converts the principal properties, say K_{123}, to the corresponding global properties with the product $K_{xyz} = T^T K_{123} T$. For the common isotropic case this reduces to $K_{xyz} = kI$, where I is the identity matrix.

13.6. General Anisotropic Material Directions

The previous section assumed that the principal material directions of an orthotropic material happened to align with the axes used to construct the part. This section is intended to show how to utilize a more general description of principal material directions. In many commercial FEA systems every element can have its own local coordinate system to define its anisotropic material directions. That places a heavy burden on the user to create such data. SW Simulation uses the more common approach of allowing the user to input a material direction associated with a single part.

To illustrate the process, for either thermal or stress analysis, a simple square part is taken to have its principal material direction inclined at an angle of 65 degrees with respect to the part's x-axis.

(1) First the square is sketched and created as a planar part with **Insert → Surface → Planar Surface**. Then a split line is constructed for use in constructing the material coordinate system, with **Insert → Curve → Split Line**.

(2) Next, create the material directions via **Insert → Reference Geometry → Coordinate System**. Select a point on the split line as the **Vertex** and the actual split line as the direction on the material's **x-axis** (first principal material direction). For a three-dimensional material direction you would have to continue and locate the y- and/or z-axis of the material.

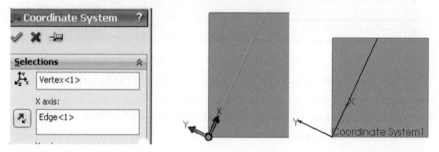

The next major step is to associate the anisotropic material properties with the coordinate system that was just constructed. Clearly, such a material has to be established as a custom material. That process was described earlier (page 94). It changes in details here in the **Material panel** where you enter the actual material property values. The new steps are:

(1) Change the default material type (Linear Elastic Isotropic) to **Linear Elastic Orthotropic**.

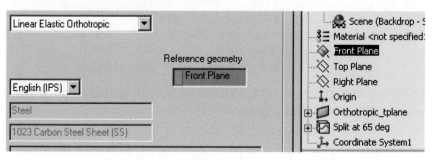

(2) You have to define the material **Reference geometry**. It usually defaults to one of the standard planes, like the Front Plane in the previous example (and the image above). In the expanded part tree, click on **Coordinate System 1** to change the default setting.

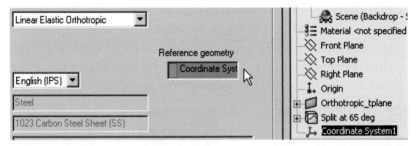

Then you enter all the anisotropic data for this material. In the thermal case they are currently restricted to the orthotropic thermal conductivities; just type in those values. Here, the material y-conductivity is ten times its x-conductivity ($ky = 10\,kx$). For this two-dimensional example there is no conduction normal to the part.

These properties define the conductivity matrix only in the local material directions. The element conduction (thermal stiffness) matrix is defined in the global coordinate system. Therefore, the local conductivity matrix must be transformed (rotated) 65 degrees to the global axes [2]. That transformation is

$$\boldsymbol{k}(\theta)_{Global} = \boldsymbol{T}(\theta)^T \boldsymbol{k}_{Local} \boldsymbol{T}(\theta),$$

where the above orthotropic conductivity matrix and 2D coordinate transformations are:

$$\boldsymbol{k}_{Local} = \begin{bmatrix} k_{11} & k_{12} \\ k_{21} & k_{22} \end{bmatrix} = \begin{bmatrix} 0.007 & 0 \\ 0 & 0.7 \end{bmatrix} \text{BTU/in} \cdot \text{s} \cdot \text{F},$$

$$\boldsymbol{T}(\theta) = \begin{bmatrix} \cos\theta & \sin\theta \\ -\sin\theta & \cos\theta \end{bmatrix} = \begin{bmatrix} 0.423 & 0.906 \\ -0.906 & 0.423 \end{bmatrix}.$$

Carrying out the matrix multiplications, the global conductivity matrix (for later use) becomes

$$\boldsymbol{k}_{Global} = \begin{bmatrix} k_{xx} & k_{xy} \\ k_{yx} & k_{yy} \end{bmatrix} = \begin{bmatrix} 0.576 & -0.266 \\ -0.266 & 0.131 \end{bmatrix} \text{BTU/in} \cdot \text{s} \cdot \text{F}.$$

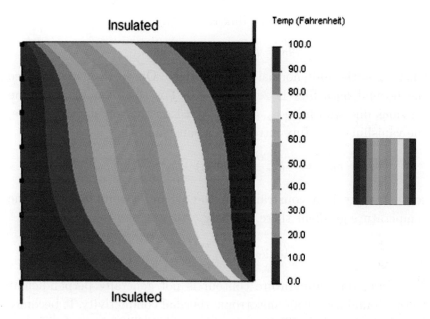

Fig. 13.14. Anisotropic (left) and isotropic material temperatures.

Here, the square part simply has constant temperatures imposed on its vertical edges (0 F on the left 100 F on the right). The part is insulated on the horizontal edges. For an isotropic material, the temperature change would be linear with $x = 0$. That is, the temperature contours of an isotropic material would be vertical and equally spaced. The thermal restraints are applied as before. The computed spatial distribution of the temperatures, in Figure 13.14, drastically differs from that expected from an isotropic material.

Of particular note is the fact that those contours do not intersect the insulated boundary at right angles, as an isotropic material always does. The reason for that is as follows. Recall that the heat flux vector is the negative product of the *global* conductivity matrix, k_{Global}, and the temperature gradient. For this two-dimensional problem the two heat flux components are:

$$q_x = -k_{xx}\frac{\partial T}{\partial x} - k_{xy}\frac{\partial T}{\partial y} \ ,$$

$$q_y = -k_{yx}\frac{\partial T}{\partial x} - k_{yy}\frac{\partial T}{\partial y} \ ,$$

while the scalar normal heat flux is

$$q_n = \vec{q} \cdot \vec{n} = q_x n_x + q_y n_y,$$

where $\vec{n}$ is the unit normal to the surface. On an insulated surface the normal heat flux is zero, $q_n = 0$. For this example, and the previous one, you have $n_x = 0$, $n_y = \pm 1$ on the top surface. There, the vanishing heat flux introduces a constraint that

$$q_y(x, y_{top}) = -k_{yx}\frac{\partial T}{\partial x}(x, y_{top}) - k_{yy}\frac{\partial T}{\partial y}(x, y_{top}) = 0,$$

so that there is a constraint on the x- and y-components of the temperature gradient there. Namely,

$$\frac{\partial T}{\partial y}(x, y_{top}) = \frac{k_{yx}}{k_{yy}}\frac{\partial T}{\partial x}(x, y_{top}).$$

Therefore, the temperature contour is not generally perpendicular to the boundary for an anisotropic thermal conductivity. It becomes perpendicular when $\partial T/\partial y = 0$ which means when $k_{yx} = 0$. That is always true for an isotropic material and for a globally orthotropic material, like the one in the previous Carslaw example. In both those cases, $\partial T/\partial y = \partial T/\partial n = 0$ and the temperature contours were normal to the top edge.

For an isotropic material with the above essential boundary conditions, the heat flux would be a constant at all points. The heat flux vectors would have only an x-component since the top and bottom edges are insulated. The anisotropic behavior of the heat flux is quite different, as seen in Figure 13.15 where the magnitude of heat flux varies widely.

The integral of the normal heat flux on a boundary is the thermal reaction needed to maintain the applied temperature. The graphs of the left and right edges are in Figure 13.16. By inspection of the last two graphs, it is clear that the integral of the heat flux (area under the curves) is the same. Therefore, the reaction heat flow in equals the heat flow out. Such thermal equilibrium must be satisfied by all materials, anisotropic or not.

Also, Figure 13.17 shows the heat flux vectors are seen to be generally inclined and indicate corner singularities in contrast with

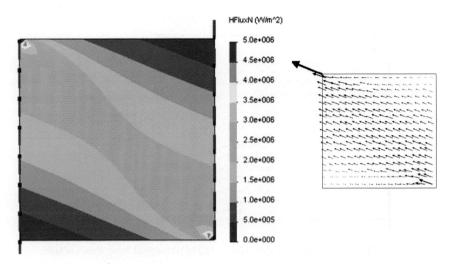

Fig. 13.15. Heat flux values for anisotropic material.

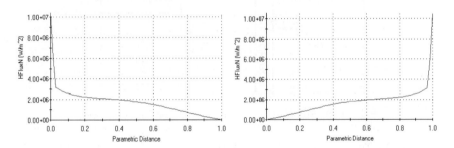

Fig. 13.16. Normal heat flow into (left) and out of the anisotropic part.

the isotropic heat flux. Of course, the heat flux vectors are still parallel to insulated surfaces since no heat flows across insulated boundaries.

13.7. Analysis of a Block with a Cylindrical Hole

A vertical square steel plate (Figure 13.18) is 30.48 cm on each side, has a 1.27 cm radius center hole, and is 10 cm thick. The steel is measured to have a thermal conductivity of 26 W/mC. The center hole carries a hot fluid and the left and right faces of the square plate have natural convection to air at 21 C and a convection coefficient

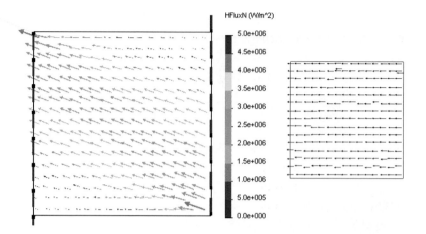

Fig. 13.17.　Anisotropic and isotropic (right) material heat flux vectors.

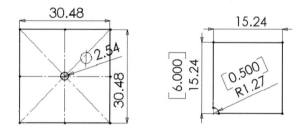

Fig. 13.18.　Identifying two planes of analysis symmetry.

of $409\,\mathrm{W/m^2C}$. The top and bottom faces of the plate are insulated. The fluid in the center hole enters at about 449 C, exits at 349 C and is assumes to vary linear along the hole. You want to estimate the heat input necessary to maintain the center hole temperatures.

The problem geometry and material have one-eighth symmetry, but the restraints and thermal load have only one-fourth symmetry. Thus the best you can do from an efficiency point of view is to model one of the 90 degree segments. Note that by "cutting" the part with two symmetry planes, you will have to assign proper boundary conditions on those two planes to account for the removed material.

Before beginning the following finite element analysis you should estimate the temperature results and/or attempted to bound them.

For a plane wall with a known inside temperature on one side and convection on the other the exact temperature solution is linear through the wall. The 1D analytic solution for a constant thickness wall estimation gives the temperature of the convection surface as

$$T_s = \frac{h_{air} T_{air} + T_{wall} k/L}{k/L + h_{air}},$$

where L is the thickness of the conduction path. The temperature along a line of symmetry can often be modeled with a 1D model that has the same end conditions as the symmetry line. Here those end conditions are the same and mainly their lengths vary. The average temperature along the inner surface is 399 C. The lower length is $L_0 = 0.1524$ m. Therefore, the estimated outside wall temperatures along the bottom edge are

$$T_0 = \frac{409 \text{ W/m}^2\text{C} * 21 \text{ C} + 399 \text{ C} * 26 \text{ W/mC}/0.1524 \text{ m}}{26 \text{ W/mC}/0.1524 \text{ m} + 409 \text{ W/m}^2\text{C}} = 132 \text{ C}.$$

The vertical symmetry line and top insulation line combine for a length of $L_1 = 0.2921$ m. Likewise, the top end point at the convection surface is estimated to have a lower value of $T_1 = 89$ C. These two estimates mean we expect the temperature on the convection surface to decrease from bottom to top points. You can also estimate the heat flow through a 1D wall there (assuming parallel heat flux vectors) as $q_0 = K(T_{wall} - T_{air})/L_0$, which gives heat flux estimates of $q_0 = 18,900$ W/m^2, and $q_1 = 9,865$ W/m^2 at the same lower and upper points of the convection surface. Averaging those two heat flux and multiplying by the convection surface area give a total heat outflow estimate of $2.17e4$ W.

You should anticipate some visual results that should appear in the post-processing. The temperature contours should be parallel to each surface with a given constant temperature (the central hole), and they should be perpendicular to any insulated surface (the top and bottom faces) and any symmetry plane. Neither case should occur at a convection boundary, except for the two special extreme cases of $h = 0$ so $T_s = T_{wall}$ and $h = \infty$ (or $h \gg k/L$) which gives $T_s = h_{air}$. Those two special conditions can exist, but they usually occur because of user data errors. Finally, the temperature and heat flux contours should be smooth. Wiggles in a contour usually mean that the mesh is too crude there. If wiggles occur in an important region the mesh should be refined there and the analysis repeated. You can also visualize some of the heat flux vector results. First, they should be parallel to any insulated surface (or symmetry condition).

From the previous thermal studies, recall that SW Simulation allows the specification of constant temperatures on surfaces. However, it does not currently allow user input of linear temperature variations required here. Thus, you could get an approximate result using the average temperature of the whole. That would be most economically done with a planar model. To get the required result, you need a way trick the simulation into applying a linear temperature essential boundary condition. That can be done by covering a surface with a false layer of highly conductive material. When the false layer is included in the study it will tend to yield a linear temperature distribution between the temperatures at its two ends. Bonding such a layer to the actual part might yield the required boundary condition. To accomplish that, the true block part was extruded as the actual material, and the actual hole was extruded as a second false material (for tricking the solver). They were combined in an assembly and meshed together as shown in Figure 13.19.

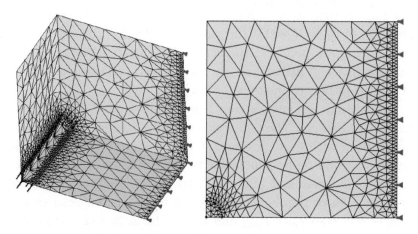

Fig. 13.19.　Block with a heated hole and a false material within the hole.

In the **SW Simulation Manager panel:**

(1) Right click on the **Simulation** → **New Study**. Set the **Study name**, thermal **Analysis type**.
(2) Under the **part name** → **Apply Material to All**. A review of the **Material panel** standard materials yields no match. Thus, a custom material input is required. This procedure has changed with the 2010 SW release (as outlined on page 94). You are now required to pick and existing material (hopefully similar), **copy** it from the standard material tree, **paste** it into the custom material region, **edit** the custom values.
(3) Here, you type in 25.95 W/m-C for the **Thermal conductivity** value, click **Apply** and **Close**. Then slow double click on the name and change it to "Thermal duct". In a similar fashion, the false material was assigned a false conductivity that was 10,000 times higher than the block thermal conductivity.

Apply the only "*essential boundary condition*" (the known linear temperature distribution). In the SW Simulation **Manager menu:**

(1) **Thermal Loads** → **Temperature** opens the **Temperature panel**.
(2) Set the **Temperature** to the inlet value of 499 C. Pick the inlet area of the false material.

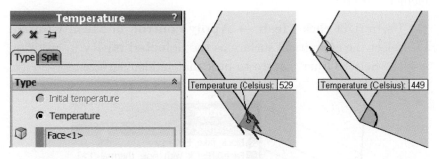

(3) Likewise, set the outlet temperature at the other end of the false material (above). This is the trick that you hope will cause the cylindrical surface to have the desired linear temperature variation.

Invoke the right side free convection as the only loading condition. In the **Manager menu**:

(1) Use **Thermal Loads** → **Convection** to open the **Convection panel**. Pick the flat convection face as the **Selected Entity**. In **Convection Parameters** set the convection coefficient, $h = 409\,\mathrm{W}/(\mathrm{m}^2\,\mathrm{C})$.

(2) Set the air **temperature** to 294 K (about 70 F).

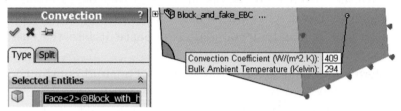

The insulated surfaces, which correspond to the top plane and the symmetry planes, require no action. That is because in any finite element thermal analysis that state (of zero heat flux) is a *"natural boundary condition"*. That is, it occurs automatically unless you actively prescribe a different condition on a boundary. This also means that the front and back of the extruded part (i.e., the "top and bottom" of your shell) are automatically insulated.

The central hole is so small that you should expect to hàve high temperature gradients there and plan ahead to assure smaller elements there:

(1) Use a right click **Mesh** → **Apply control**. In **Mesh Control** select the cylindrical surface as the **Selected entity** by utilizing the Select Other feature to pick that hidden surface.

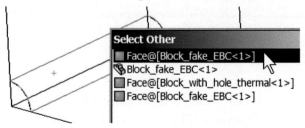

(2) Observe the default element size and reduce its value in the **Control Parameter** to 0.08 inch, click **OK**. In the **Manager**

menu right click **Mesh** → **Create**. In the **Mesh panel** click **OK** for the mesh generation shown earlier in Figure 13.19.

Start the temperature solution with right click on the **Name** → **Run**. Usually you get a solution completed message. Here, the false material might cause a numerical conditioning problem. Sometimes the fast iterative solver might fail. If that happens you need to change to the sparse direct solver (under **Tools** → **Options** or **Run** → **Properties**) which is slower, but more robust.

Begin the results review with a temperature plot to access if the boundary condition trick worked. In the **Manager menu** under **Thermal**:

(1) Double click on **Temperature** → **Plot 1**. The default contour plot of temperatures would appear as a smoothed (Gouraud) color image. Usually a stepped shaded image gives a better hint of a bad mesh.

(2) To create one right click in the graphics window, **Edit Definition** → **Thermal Plot** → **Display**. Set **Units** to Celsius and change **Fringe type** to discrete filled, click **OK**.

(3) Right click in the graphics window, select **Chart Options**. In the **Color Map** panel pick 8 **colors** of thin **Width** and 2 **Decimal** places, click **OK**. Such a typical temperature plot is seen in Figure 13.20 on the left side. The right side of the figure

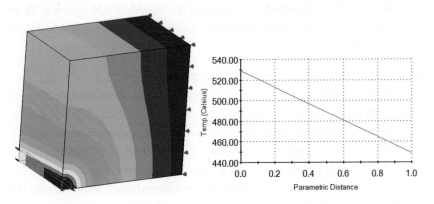

Fig. 13.20. Temperature level along the true hole edge.

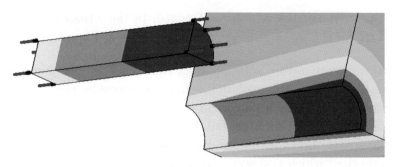

Fig. 13.21. Exploded temperatures of the true and false hole material.

shows a List Selected graph along the edge of the true hole, from the inlet to the outlet side.

(4) Activate an exploded view of the temperatures to further verify that the trick has accomplished the desired linear temperature essential boundary condition along the true hole surface (Figure 13.21).

You can also obtain graphs of selected results along the boundary of the part. To obtain a temperature graph:

(1) Right click in the graphics area of a temperature plot, pick **List Selected** . . .
(2) Pick the desired edge (lower straight symmetry line) as the **Selected items**. Click on **Update**.
(3) To see a graph of the temperature along that edge select **Plot**. That graph indicates a bad mesh if the graph is not smooth. Here (Figure 13.22) it seems smooth, but it has a very sharp gradient at one end.

Remember that the 1D approximation would give you a straight line between the max and min temperatures, so the actual temperature graph gives you some feel for how much to trust such a 1D estimate.

The temperature distribution on the free convection surface is usually of specific interest. It also can sometimes be compared to known solution chosen to try to estimate a correct result, as is done here with a 1D estimate. Therefore it is desirable to supplement the above probe operation with an edge temperature summary and a

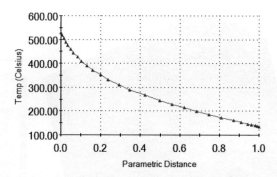

Fig. 13.22. Graph of temperatures along lower symmetry line.

graph. That is accomplished by repeating the last set of operations, but selecting the insulated edge as the selected item.

The top and bottom edges were found to be a 113 and 110 C, respectively. The 1D hand calculated edge temperature range along the convection surface ranged from 132 C down to 88 C and average to 110 C. That agrees reasonably well with the computed range.

The heat flux is a vector quantity defined by Fourier's law: $q = -K\nabla T$. Thus, it is best displayed as a vector plot for two-dimensional problems. However, for three-dimensional parts the vector views can be confusing.

(1) Right click in the graphics window, **Edit Definition** → **Thermal Plot** → **Display**. Set **Units** to W/m^2.
(2) Pick **Component** resultant heat flux. The magnitudes are given in Figure 13.23.

Like with the temperatures, one can obtain summary results for the heat flow through a surface of the block. Here, the heat enters along the hole and exits at the convection surface. To obtain a heat flux summary:

(1) Right click in the graphics area of a heat flux plot, pick **List Selected** . . .
(2) Pick the desired cylindrical hole surface the **Selected items**.
(3) Click on **Update**. The summary of the normal heat flux values along that path appear in the list at each node on that path. The

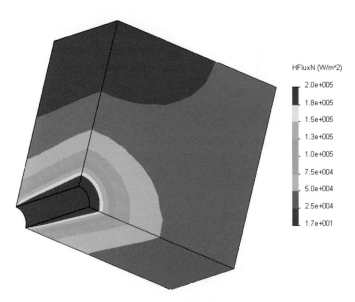

HFluxN (W/m^2)

2.0e+005

1.8e+005

1.5e+005

1.3e+005

1.0e+005

7.5e+004

5.0e+004

2.5e+004

1.7e+001

Fig. 13.23.   Heat flux (nodal) magnitudes from linear temperature hole.

Avg, **Max**, and **Min** heat flux magnitudes appear in the **Value** column.

(4) Note that the **Value** column also contains the **Total Heat Flow** as about 500 W (positive in, negative out) across the selected surface. It is the integral of the normal heat fluxes over the curved surface area selected. In other words, it is the heat flow in necessary to maintain the given temperature.

The above heat flow into this system (given above) should be equal and opposite to the heat flow going out at the convection face (since there were no internal heat generation rate data). If they do not reasonably agree then the mesh should be revised. Such differences occur since they are calculated from the gradients of an approximate temperature solution. The temperatures are always more accurate that the heat flux, but you need to have acceptable accuracy for both. Repeating the above procedure for that convection surface gives the total heat flow (Figure 13.24) as about −505 W. That is an acceptable difference of only 1%. However, a refined mesh could easily reduce the error in this equilibrium check of the thermal reactions.

Summary		
	Value	
Total Heat Flow	500.12	W
Avg	1.0305e+008	W/m^2
Max	1.4767e+008	W/m^2
Min	5.1479e+007	W/m^2
RMS	1.0413e+008	W/m^2

Summary		
	Value	
Total Heat Flow	-505.05	W
Avg	33239	W/m^2
Max	35359	W/m^2
Min	31081	W/m^2
RMS	33272	W/m^2

Fig. 13.24. Heat flow balance at the temperature and convection surfaces.

A heat flow in or out of the system must occur at every specified temperature node and at any convection nodes. If your finite element system provides those data it is good practice to review them. SW Simulation does allow you to recover those data, as shown above, but if it did not basic engineering would give you an estimate of the total heat flow (per unit length assumed here for the thickness into the page). Such a validation check is important since it is not unusual for a user to enter incorrect values for the K and h values. The ratio of those values is important in convection calculations.

From the vector plot of heat flux in Figure 13.23 you see that at both the inner cylindrical surface and the outer insulated surface that the flow is basically normal to the surface. Integrating the normal heat flux passing through either surface gives the total heat flow lost. Your computer model was only 1/4 of the total domain. Therefore, the true heat loss is about $Q_{total} = 2,000$ Watts.

13.8. Crossing Pipes Analysis

Chapter 7 of Ref. [7] outlines a steady state thermal analysis of an assumed pipeline junction. Here alternate points of view and additional post-processing features will be presented. The first difference is to recognize that the geometry, material properties and boundary conditions have a plane of symmetry. Therefore, a half model can be employed. That lets the full mesh be efficiently applied to non-redundant results. The half model is seen in Figure 13.25, where each pipe hot end surface has been color coded light red, the original interior is green, and the material exposed on the symmetry plane

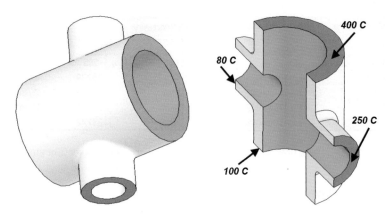

Fig. 13.25.   External and section view of crossing pipes.

is in yellow. This is one of those problems where the temperature solution depends only on the geometry and is independent of the material used. Of course, the heat flux vectors and the thermal reactions will always depend on the thermal conductivity, k.

The larger junction has inner and outer diameters of 3.5 and 5.0 inches, respectively. The smaller junction has inner and outer diameters of 1.0 and 2.5 inches, respectively. The overall length of each section is about 8 inches. The material is brass.

Most experimental properties are known only to two or three significant figures. However, the library table values can be misleading because the properties sometimes were measured in a different set of units and multiplied by a conversion factor and incorrectly displayed to 7 or 8 significant figures. The SW tabulated thermal conductivity of brass is displayed in its experimentally measured units as $k = 110$ W/m-K, but had you used English units it would be converted and displayed to 8 significant figures instead of a realistic value of about $k = 1.47e$-3 BTU/in-s-F.

This pipe junction has four sets of restraints, or "essential boundary conditions", where the temperature is specified at each pipe end ring surface (Figure 13.25 right). The first is applied by selecting the hottest end area and assigning it a given value of 400 C. The other three ends are treated in a similar way. The application of the first restraint is illustrated in Figure 13.26. That figure also

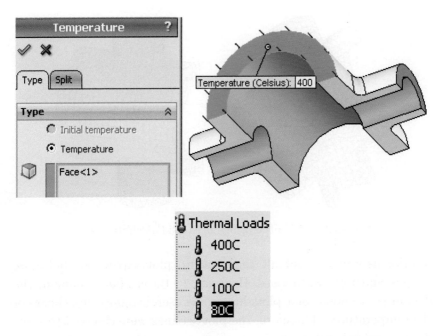

Fig. 13.26. Inlet temperature at larger pipe, and assigned restraint names.

shows that the default restraint names have been replaced with more meaningful ones. That practice often saves time later when a problem has to be reviewed.

The boundary condition on the (yellow) symmetry plane must be introduced to account for the removed material. Since it is a plane of symmetry it acts as a perfect insulator. That is, there is no heat flow normal to the plane ($q_n = 0$). That is the "natural boundary condition" in a finite element analysis and is automatically satisfied. The original interior surface also has not yet been specifically addressed. Neither has the remaining exterior (gray) surface. They both also default to insulated (or adiabatic) surfaces having no heat flow across them. That is probably not realistic and convection conditions there will be considered later in the appendix. All that remains is to generate a mesh, compute the temperatures and post-process them.

The automatic mesh generator does a good job. The inner and outer views of the default mess are in Figure 13.27. When the

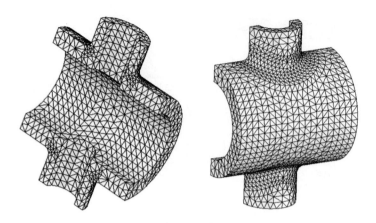

Fig. 13.27. Default crossing-pipes solid quadratic mesh.

solution is run the default temperature plot shows two hot pipe regions and two cooler ones. However, in the author's opinion, the default continuous color plots hide some useful engineering checks of the temperatures. Therefore, the plot settings were changed to show discrete color bands.

The inner and outer surface temperature displays are seen in Figure 13.28. They make it a little easier to check that the contours are perpendicular to the flat symmetry plane, and nearly parallel to surfaces having constant temperatures.

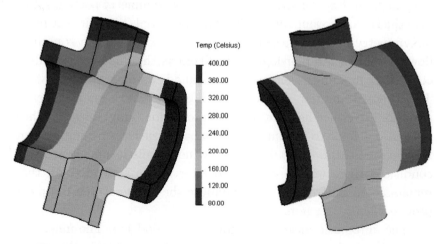

Fig. 13.28. Half model internal and external conduction temperature.

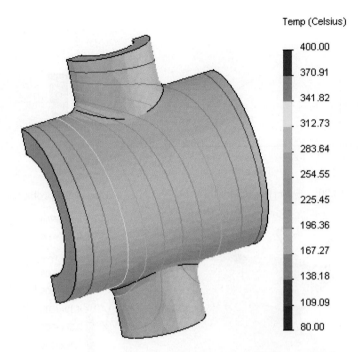

Temp (Celsius)

400.00

370.91

341.82

312.73

283.64

254.55

225.45

196.36

167.27

138.18

109.09

80.00

Fig. 13.29. Line contour exterior temperature result for conduction only.

Another alternative is to change the settings to line contours to obtain the results of Figure 13.29. You can also change the number of color segments to reduce or increase the number of contours lines displayed in either mode. It is common, and desirable, to make technical reports more specific by graphing the results along selected lines or curves. That is done by utilizing the **List Selected** option after the temperature has been plotted. Then you select an edge, of model split line, pick update and then plot.

The selected line, at an inner edge adjacent to the hottest inlet, is seen in Figure 13.30 along with the resulting temperature graph versus the non-dimensional position. Since there is no convection, heat generation, or non-zero heat flux conditions the temperature results and contours are the same for all materials. However, the heat flux does change magnitude with different materials.

The heat flux magnitude contours are given in Figure 13.31. The contour lines are less smooth than the previous ones for the

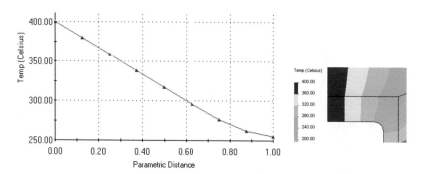

Fig. 13.30. Selecting model lines for graphing temperature probe lists.

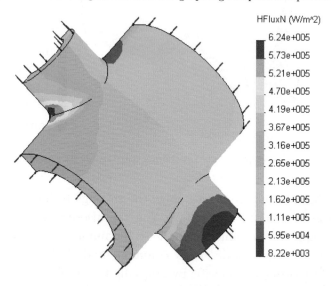

Fig. 13.31. Conducting crossing pipes heat flux contours.

temperature because the heat flux is always less accurate than the temperatures. Had these contours shown larger wiggles it would be a signal that a finer mesh should be utilized. Since the heat flux is a vector quantity it should also be plotted as a vector. The resulting color vector plot is shown in Figure 13.32 (and in black and white after changes with **Vector Plot Options**). The heat flux vectors show that heat flows into the junction at the 400 C surface and out at the 80 and 100 C ends. The 250 C end has some inflow and some heat outflow.

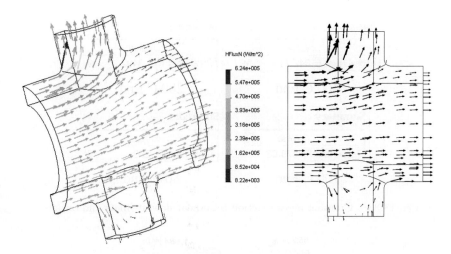

Fig. 13.32.   Conducting pipes internal heat flux vector distribution.

Here you can also have SW Simulation compute the thermal reactions necessary to maintain the specified temperatures. To do that, use the **List Selected** option and select each of the four pipe ends in order. At each one, you pick **update** to list and sum all the individual nodal heat flux values. Figure 13.33 shows those four total heat flows. They show that a total of about 850 W of power in (positive) at the two hottest surfaces and out (negative) the two coolest pipe ends. Since a half symmetry model was used here, that figure needs to be doubled to determine the required input power of about 1,700 W to maintain the specified temperature restraints. If the current example were changed to steel with a conductivity of about $k = 51.9$ W/m-K then the heat flux magnitudes would drop by about a factor of two while the temperature would be unchanged. Figure 13.34 shows the system heat flow reactions.

The neglected convection conditions or known heat flow on the boundaries will change the temperature distribution in the original problem. To illustrate this point, assume the outer surface convects to air at a temperature of 30 C (303 K) with a convection coefficient of about $h = 5$ W/m². Also let the pipe interiors convect to oil at 70 C (343 K) with an assumed convection coefficient of about 600 W/m². You simply have to apply two convection conditions (seen in Figure 13.35), and re-compute the solution.

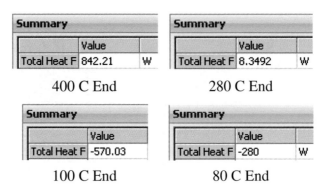

Summary		
	Value	
Total Heat F	842.21	W

400 C End

Summary		
	Value	
Total Heat F	8.3492	W

280 C End

Summary		
	Value	
Total Heat F	-570.03	

100 C End

Summary		
	Value	
Total Heat F	-280	W

80 C End

Fig. 13.33.　Crossing pipes reaction total heat flow at inlets and outlets.

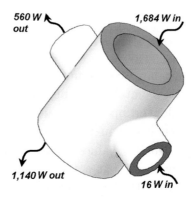

Fig. 13.34.　Pipe heat flow equilibrium check.

As expected, the convection effects cause a greater heat loss and thus higher temperature gradients (and thus heat flux) near the inlet regions. That can be seen by comparing the original temperatures in Figure 13.28 to the revised ones in Figure 13.36. The original graph of temperatures along the interior edge in Figure 13.30 shows less temperature drop than the new graph in Figure 13.37. There are steeper temperature gradients than seen in the original problem. The heat flux vectors, including convection, are given in Figure 13.38.

When the integrals of the normal heat flux were summed, using the values from the above mesh, the thermal reactions were out of balance by about 40%. Therefore, the mesh was refined and the solution re-computed. The temperatures change relatively little, but the heat flux was revealed in more detail. The heat flux from the

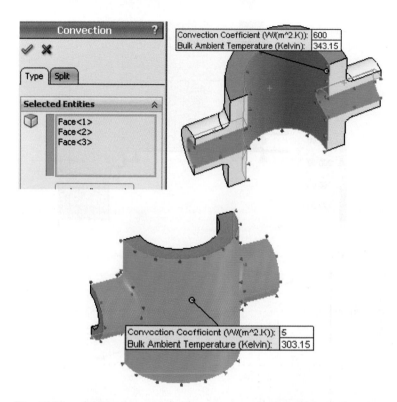

Fig. 13.35. Adding internal (left) and external convection to the pipes.

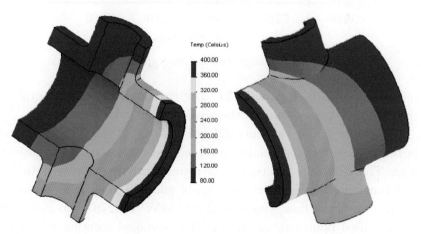

Fig. 13.36. Rapid temperature reduction due to convection.

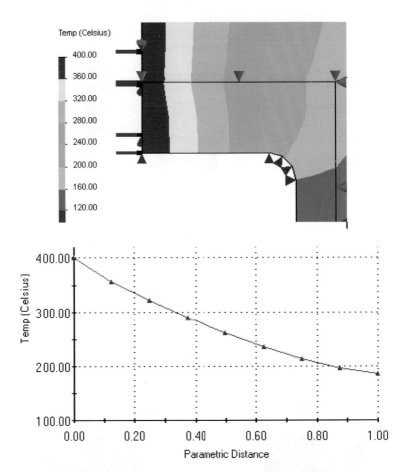

Fig. 13.37.　Revised edge temperature graph along inner edge.

refined mesh is in Figure 13.39. The reactions are the integral of those surface values (sum of the areas of each color region).

The prior larger elements did not pick up all the fine variations in heat flux values, especially the small red regions near the hottest inlet. Therefore, the reactions were significantly in error, despite the fact that the temperatures were quite good. If you again compute the reaction heat flow, you need $2 * 2{,}138\,\text{W} = 4{,}276\,\text{W}$ to maintain the specified boundary conditions. The summed reaction heat flow reactions for the six regions are given in Figure 3.40. These reactions are quite different from those in Figure 13.34.

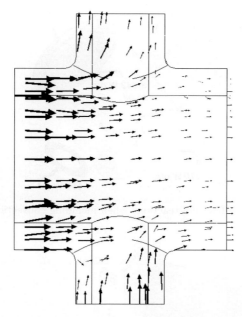

Fig. 13.38. Internal heat flux vectors with convection effects.

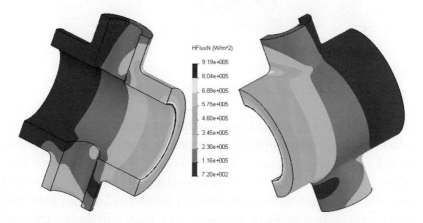

Fig. 13.39. Heat flux values from a refined mesh.

13.9. Thermal Analysis of a Ram Block

A high pressure flow line has a safety shut-off section that allows an oval shape ram to enter from two sides and seal off the flow. The component shown in Figure 13.41 is made of C276 steel. The center

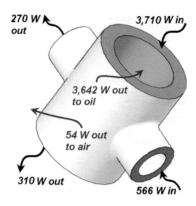

Fig. 13.40.   Thermal reactions when including convection losses.

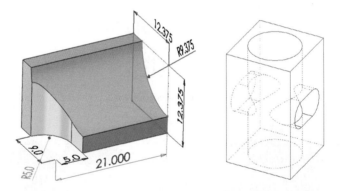

Fig. 13.41.   Solid ram block, its eighth-symmetric corner.

cylindrical internal passage carried a fluid at 400 F, while the exterior surface was cooled by natural convection by surrounding cool water (at 45 F). The convection coefficient there is typical for water, $h = 2.89e\text{-}4\,\text{BTU/s in}^2\,\text{F}$. A uniform temperature rise of an unrestrained body will not cause thermal stresses to develop. However, here you should expect non-uniform temperatures that will cause thermal stresses. Here you will get the non-uniform temperature distribution, using a conductivity of $k = 1.736e\text{-}4\,\text{BTU/s in F}$, but delay the thermal stress load case until later. (Review the SW Simulation tutorial on thermal stresses.)

The oval ram access hole will be assumed to be filed with another solid surrounded by an insulating material. Later, you could repeat this study in an assembly that has those materials. That would give a more accurate temperature distribution around the oval opening (which is the main location of concern).

Before beginning a 3D study it is wise to estimate some aspect of the answer in advance. The estimate might be analytic, 1D or 2D finite element, or a combination of those. For plane walls the exact temperature varies linearly through the wall and linear finite elements can give an exact solution, with a single element through each material. But this part is closer to a thick wall cylinder, which generally has a logarithmic temperature distribution. Thus, a typical 1D finite element solution would need more than one axisymmetric element to get a reasonable approximate solution.

There is an exact solution for a thick cylinder with the inside temperature given and convection on the outer surface. Here you could use that to estimate the unknown part temperature at the convecting water surface. You could get bounds by using the minimum and maximum (corner) wall thickness. Jiji [6] gives the analytic solution for the temperature through the cylinder. Let a and b denote the inner and outer radii, respectively. For a conductivity of k and convection to water at a temperature of T_w with a convection coefficient of h_w the logarithmic temperature distribution is:

$$T(r) = \frac{T_a - \ln(r/a)(T_a - T_w)}{k/bh_w + \ln(b/a)}.$$

The radial heat flow (W/s) per unit length of the cylinder is constant but the heat flux, q_r, is not, because the cylindrical area of flux flow, $A = 2\pi r L$, is not constant at each radial position. The heat flux (W/s in^2) is

$$q_r = \frac{k}{r} \frac{(T_a - T_w)}{k/bh_w + \ln(b/a)}.$$

To estimate the body temperature at the surrounding water surface simply set $r = b$. From the given data $a = 9.375$ in, for this part 12.375 in $\leq b \leq 14.5$ in so say $b = 13.4$ in. Here $k = 1.736e$-4 BTU/s in F, $T_w = 45$ F, $h = 2.89e$-4 BTU/s in^2 F so that $T(b) = T_b = 84.6$ F is a reasonable average estimate, and rising to about 98 F at the thinnest section. The heat flux is about $q_r = 0.020$ BTU/s in^2 at the inner radius, and drops to about 0.015 BTU/s in^2 at the outer radius.

The part has one-eighth symmetry, so that relatively simple geometry will be employed in the thermal study:

(1) Construct the solid body and give it a name, say Block_thermal.
(2) **New → Part → OK**. Right click on **Simulation → New study** to get the **Study panel**. Enter a **Study name**, pick thermal as the **Analysis type**, click **OK**.
(3) **Apply/Edit Material**: use the custom material properties for C276 steel that were defined earlier (thermal conductivity, $K = 1.736e\text{-}4$ BTU/in s F).

Only the inner circular surface needs a thermal restraint:

(1) **Thermal Loads → Temperature** opens the **Temperature panel**.
(2) There pick the cylindrical face as the **Selected entity**, enter 400 F as the **Temperature**, click **OK**.

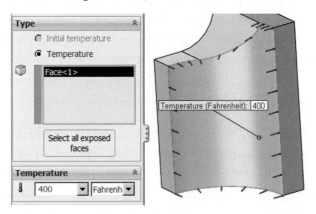

The two flat outer walls have convection to the surrounding water. Apply it via:

(1) **Thermal Loads → Convection** to open the **Convection panel**.
(2) There pick the two faces as the **Selected Entities**. Enter $2.89e\text{-}4$ BTU/s in^2 F for h, and 45 F for the **temperature** under the **Convection Parameters**, click **OK**.

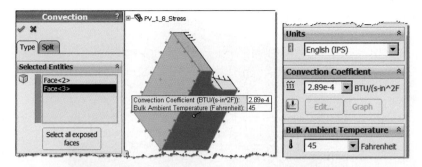

The remaining five surfaces are insulated. They include the three flat rectangular symmetry planes, and by assumptions the ram oval access surface and the top most surface. An insulated surface (zero normal heat flow) is a natural boundary condition in finite element formulations. They require no action or input.

Were the oval hole not present you would expect the temperature contours to be very close to those in a thick cylinder. Handbook 1D solutions are available for thick wall cylinder. That relatively uniform temperature distribution will be disrupted by the (insulated) oval hole. Therefore, you should control the mesh:

(1) Right click on **Mesh → Apply Control** to open the **Mesh Control panel**. There pick all the edge curves on the oval surface to be the **Selected Entities**, just to be safe.

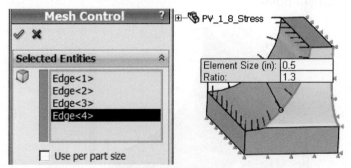

(2) Under **Control Parameters** reduce the element **size** from the default displayed value to 0.5 inch, click **OK**. The resulting mesh given in Figure 13.42 is reasonable fine.

Pick **Study name → Run** to generate the solution.

The only plots and probes are under the thermal report. Start with the default temperature plot:

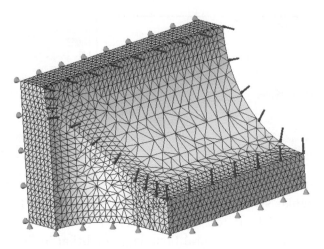

Fig. 13.42.   Initial solid mesh with thermal symbols.

(1) **Results → Define Thermal Plot → Temperature** gives the
    surface temperature plots. They will include the extreme values
    since no internal volumetric heat generation was active here.

(2) **Rotate** the temperature and search for the region of closest
    contour lines (highest temperature gradient) since that region
    would cause the highest thermal stresses. As expected, that
    occurs on the internal intersection line between the oval and the
    cylindrical holes.

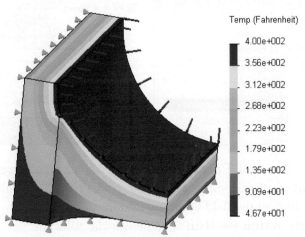

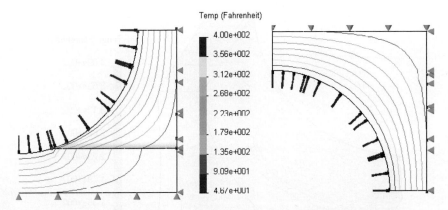

Fig. 13.43.   Temperature at middle intersection edge, and top (right).

(3) Examine that region in more detail in the bottom model view of
Figure 13.43. Pick **Settings → Fringe Options → Line**.

Even though the highest temperature will be on the surfaces, it may
be useful to also plot an interior section so as to see how close most
of the body would be to the 1D axisymmetric analytic solution:

(1) Right click in the graphics area, and pick **Section clipping**.
(2) Select the **Top plane**, **Show uncut portion** and move the
slider to see various sections.

As expected, Figure 13.44 shows that the internal contours are almost
nested cylindrical isosurfaces like (the logarithmic distribution) of a
thick walled cylinder.

Regions of high thermal gradient, and thus high heat flux, may
be missed with a smoothed temperature contour plot. Thus, also
display them with:

(1) Right click in the graphics area, **Edit Definition**. Select resultant heat flux (**HFLUXN**) and the desired **units**.
(2) Right click for **Settings → Fringe Options → Discrete**
(Figure 13.45).

The magnitude display of Figure 13.45 confirms the location of
concern for future thermal stress studies. Complete that check by

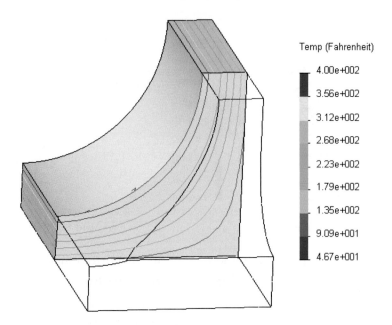

Fig. 13.44.   Nearly cylindrical temperature contours in an upper section.

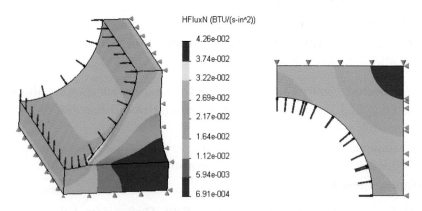

Fig. 13.45.   Magnitude of the heat flux peaks at the inner intersection.

looking at the heat flux vectors:

(1) Right click in the graphics area, **Edit Definition**. Select **Advanced → Vector Plot** click **OK**.

(2) Right click on the **Plot name → Vector Plot Options** and scale the size and number of vectors.

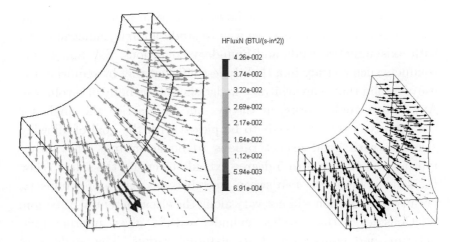

HFluxN (BTU/(s-in^2))

4.26e-002

3.74e-002

3.22e-002

2.69e-002

2.17e-002

1.64e-002

1.12e-002

5.94e-003

6.91e-004

Fig. 13.46.   Wireframe view with heat flux vectors.

The display in Figure 13.46 (above) yields a final verification of the concern about the intersection curve area having high temperature gradients.

The estimated thin wall surface temperature of about 84.4 F agrees very well with the temperature contours seen in Figure 13.43 which are about 90 F at the surface. The approximate average radial heat flux of about $1.8e\text{-}2\,\mathrm{BTU/s\,in}^2$ is also in good agreement with the analytic estimate.

If you were not aware of the analytic approximation, or know how to derive it you could find a 1D (axisymmetric) finite element solution to estimate the expected results. The summary of such a solution is based on the typical linear element described in Akin [2] for approximating radially symmetric heat problems. A single element could give an analytic estimate of the surface temperature. Such an element has a constant heat flux, so several should actually be utilized in a 1D validation check.

13.10. Axisymmetric Thermal Analysis

Many FEA systems have a separate module for axisymmetric problems defined by completely revolving a planar area about a central axis. That is, the model is described as a planar part and the hoop effects are brought in through the mathematical model. That is the most computationally effective way to study an axisymmetric body, with axisymmetric loads and boundary conditions. SW Simulation conducts such a study in a slightly less efficient way by requiring the use of a solid that is an arbitrary wedge slice of the body of revolution. Any wedge angle works, in theory, but in practice the angle for the "revolve extrude" needs to be picked with the goal of allowing elements with good aspect rations to be generated. The author typically employs a 3 to 5 degree extrude to create the solid. When the solid is built, the two sweep faces become planes of symmetry, because the solution will not vary along the circumferential direction.

Consider a thick walled cylinder with a given internal temperature and convection at its external radius. The analysis of heat transfer in a thick wall infinite cylinder is a one-dimensional problem, and the analytic solution is known [1]. The cylinder has inner and outer radii of 9.375 and 13.40 inches, respectively, and is made of aluminum alloy 1345. The inner and outer temperatures are 500 and 45 degree F, respectively, with and outer convection coefficient of $2.89e\text{-}4\ \mathrm{BTU/F\text{-}s\text{-}in}^2$. The goal is to determine the inner temperature distribution, and the required heat flow through the wall. This problem provides a chance to verify your knowledge of SW Simulation, and to illustrate some optional features. To access some optional features you must have a part open, or prepare to open the first part with **File** $\rightarrow$ **New** $\rightarrow$ **Part** $\rightarrow$ **OK**.

Begin a sketch of the cross-section by inserting the centerline of revolution at the origin. Construct a rectangle with the vertical sides located at the inner and outer radial distances. Have the bottom of the rectangle at the level of the origin and assign it an arbitrary thickness on 1 inch. With the sketch open, select **Features** $\rightarrow$ **Revolve Boss/Base**. In the **Revolve panel** select the vertical axis; specify the angle as 5 degrees behind the front plane, **OK**. The wedge section is seen in Figure 13.47.

Revolving the area behind the front plane simply makes it easier to examine the typical planar results true shape in the front view.

Next, enter SW Simulation by selecting its manager icon. In SW Simulation:

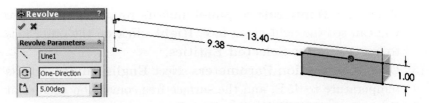

Fig. 13.47.   A wedge solid used to represent an area of revolution.

(1) Right click on the **Simulation → New Study** to open the **Study panel**.
(2) Insert a **Study name** (say Inf_cyl_thermal_wedge), select **Thermal** from the **Analysis type** list.

Begin with the essential boundary condition of the inner specified temperature:

(1) In the SW Simulation manager right click on **Thermal Loads → Temperature**. When the **Temperature panel** appears **right click** on the inner face to insert it as the **Selected Entities**.
(2) Under **Temperature** set the **units** as degrees Fahrenheit and insert a **value** of 500. Hit **Preview** to verify your assignment, click **OK**.

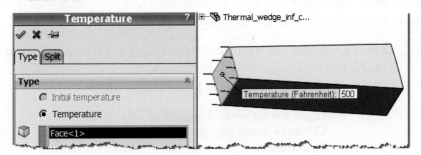

Next apply the convection conditions of the outer cylindrical (light green) face:

(1) In the SW Simulation manager right click on **Thermal Loads → Convection**.

(2) When the **Temperature panel** appears rotate the part until you can see the light green face. **Right click** on the outer face to insert it as the **Selected Entities**.

(3) Under **Convection Parameters** select **English units**, set the temperature to 45 F, and the surface free convection coefficient (**h**) to 2.89e-4 BTU/F-s-in². Click **Preview**, and then **OK**.

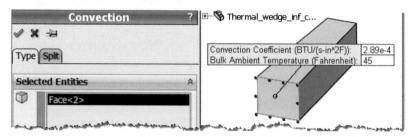

Convection is a boundary condition of the mixed or Robin type. It can approach either a specified temperature surface or an insulated surface. Usually those limits are only reached with user input errors. Also keep in mind that the remaining four surfaces of this part are automatically insulated in an FEA, unless you assign another condition.

Another very useful option is the ability to **assign names** to any item in the **construction tree**. This helps you remember your thought process when you need to come back at a later time to review your analysis or design. In the manager panel:

(1) Slow double click on **Temperature-1** under **Thermal Loads**. Enter a descriptive term, say Hot_Inside_T.

(2) Likewise, change **Convection-1** to Convect_H2O.

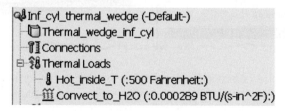

The material is in the standard materials library: Right click on the **Part name** → **Apply/Edit Material** to open the **Material**

panel. Select **SolidWorks Materials** → **Aluminum Alloys** → **1345 Alloy, OK**.

For this 1D problem only a few elements are needed in the radial direction. Thus, the default mesh will be generated:

(1) In the **manager tree**, right click **Mesh** → **Create** to open the **Mesh panel**.

(2) In the **Mesh panel** accept the default element size and transition controls, click **OK**. Do not check "Run study after meshing". You should always check the mesh first.

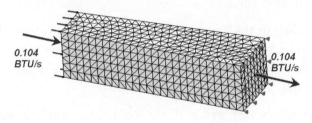

(3) Visually **inspect** the mesh. Having about 20 elements in the radial direction should be fine. The solution in the circumferential direction and in the axial direction should be constant, so the number of elements in those directions does not matter (but they do cost).

The thermal analysis is ready for solution. Right click on the **Study name** → **Run**. After the equation solver reports a successful calculation the post-processing results can be reviewed and plotted.

Begin the study review by examining the temperature distribution:

(1) In the **SW Simulation manager tree** click on **Temperature** and the double click **Plot-1**. The default plot is a continuous color contour.

(2) For an alternate format right click in the graphics area and pick **Settings**. In the **Thermal Plot** panel pick a **Discrete** for the **Fringe Type**, click **OK**. The discrete contours appear. The convection surface temperature is about 364 F. The temperature varies logarithmically with radial position in cylinders.

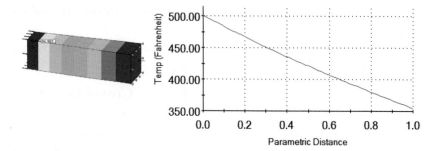

Graphs can provide more detail in selected regions. Graph the radial temperature first (above):

(1) Right click in the graphics window, pick **List Selected**.
(2) Select the **bottom edge**, click **Update** to see max, min, and average values. In **List Selected panel**, pick **Plot** to display the graph.

The graph agrees very well with the logarithmic analytic solution. A reasonable estimate could have been obtained with a single element hand solution to help validate the temperature result.

The heat flux is a vector quantity obtained from the scalar temperature. In this case, it must be in the radial direction (plot the vector form to see that) so just the values are shown here in Figure 13.48.

Integrating the heat flux over the inlet and outlet surfaces yields the respective thermal reactions of 0.104 BTU/s into the part

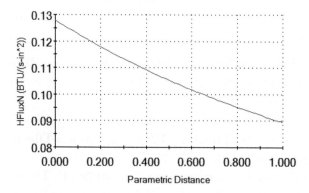

Fig. 13.48.   Radial heat flux magnitudes.

Summary

	Value	
Total Heat Flow	0.10459	BTU/s
Avg	0.12784	BTU/(s-in^2)
Max	0.12785	BTU/(s-in^2)
Min	0.12784	BTU/(s-in^2)
RMS	0.12784	BTU/(s-in^2)

Summary

	Value	
Total Heat Flow	-0.10459	BTU/s
Avg	0.089444	BTU/(s-in^2)
Max	0.089446	BTU/(s-in^2)
Min	0.089441	BTU/(s-in^2)
RMS	0.089444	BTU/(s-in^2)

Fig. 13.49. Heat flux summary checks confirm thermal reactions.

(positive) and an equal amount leaving. The constant heat flux in was 0.128 BTU/s-in^2 which drops to 0.089 BTU/s-in^2 through the larger outlet area. Of course, the area of conduction here used in finding the thermal reactions was arbitrary. Even though the area is arbitrary, the thermal reactions are equal and opposite, as shown in Figure 13.49. The point is that you can extract those data, and they are physically important in some cases.

14

Thermal Stress Analysis

14.1. Introduction

Non-uniform temperature distributions in a component cause deflections and stresses in the part. Such "thermal loads" are difficult, if not impossible, to visualize and usually need to be determined by a thermal analysis. The output temperatures from that thermal analysis can be used as input data for a stress analysis. Design tasks often involve parts made of more than one material. Sometimes the materials are bonded together and at other times they are not. Assembled parts may look like they consist of bonded parts simply because appear to be touching. However, analysis software may default that adjacent parts are not bonded or even in contact unless the user specifically establishes such relations.

14.2. Layered Beam Thermal Stress Model

To illustrate these concepts consider the common elementary physics experiment of uniformly heating two bonded beams made of materials having different coefficients of thermal expansion, α. When subject to a uniform temperature change, ΔT, the bonded beam takes on a state of constant curvature, of radius r, even though there are no externally applied forces. The unstressed length is L, and the final end deflection as δ (see Figure 14.1). Such devices are commonly used in mechanical switches that are temperature activated.

To illustrate this type of bimetallic strip, make a relatively long beam of rectangular cross-section from steel, and one with identical geometry of copper. You will save the two parts and then bring

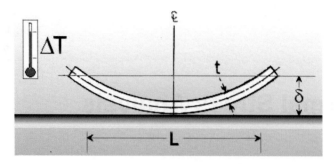

Fig. 14.1.   A bimatellic strip.

them together in an assembly. Let the length be $L = 1,000\,\text{mm}$, width $w = 80\,\text{mm}$, and depth $d = 10\,\text{mm}$ for each material. When assembled, the two members will have a total thickness of $t = 2d$ and two planes of symmetry of the geometry, material properties, and the uniform temperature change. Thus, you will use a one-quarter symmetry model. One symmetry plane will be at the middle of the length and transverse to it. The second will pass through the long axis of the beam and divide it into two equal sizes. Therefore, each of the material parts will be 500 mm by 40 mm by 10 mm.

Construct the steel first beam: Right click **Front** → **Edit Sketch** → **Insert rectangle**. Set the width to 40 mm and the depth to 10 mm. Use **Extruded Boss/Base** → **Extrude panel**. Pick **Blind** and a distance of **D1** = 500 mm, click **OK**. Now that you have built the first beam body, assign it the material property of alloy steel. Then save the part with **File** → **Save as** filename steel_beam.sldprt.

Open a new (second) part to construct the copper beam portion. Repeat the processes above except select copper as the material and save it with the filename copper_beam. Use **Window** → **Tile Horizontally** to see both parts.

Now open a new assembly:

(1) Use **File** → **New** → **Assembly** → **OK**. In the **Open Document** panel click on the steel beam in its window and drag it to the assembly window. As the first part it will be fixed in the view and other parts will move relative to it.

(2) Click on the copper beam part and drag it into the assembly window.

Next you need to carry out the geometric mating of the different materials so that the surfaces that may be bonded in the finite element analysis are touching in the assembly:

(1) Select the **Mate** (paper clip) icon. In the **Mate panel** pick **concentric** as the **Standard Mate**.
(2) Select the long bottom edge of the copper and the long top edge of the steel as the **Mate selections**.

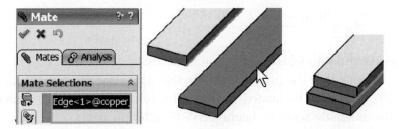

(3) In the **Mate panel** pick **concentric** as the **Standard Mate**.
(4) Select the short bottom front edge of the copper and the short top front edge of the steel as the **Mate selections**, click **OK**. Finally, **File** → **Save as** file name copper_steel.sldasm.

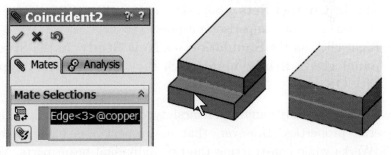

Before leaving the assembly, note that the copper part was placed on top of the steel part. That should give you a hint on the deflection directions that will be computed in the SW Simulation study.

Before building a full computer model you should try to estimate what to expect as an answer. As the assembly was built, the copper was on top. Since it has a higher coefficient of thermal expansion the top will get longer than the bottom. If they are bonded then beam should bend away from the copper. If bonded there will be a common axial extension, which is less than the free thermal expansion of the copper, but more than the free expansion of the steel. An equal and opposite set of internal bonding forces are developed at the common (equilibrium) extension. The classic handbook "Roark's Formulas for Stress and Strain" [16, 17] gives the beam theory solution for the deflections and stresses for a bimetallic strip (neglecting Poisson's ratio included here). The maximum deflection for equal thickness material layers reduces to:

$$\delta_y = \frac{6(\alpha_s - \alpha_c)(T - T_0)L^2}{K_1 t^2}, \quad K_1 = 14 + \frac{E_s}{E_c} + \frac{E_c}{E_s}.$$

The given data yields a deflection estimate of $\delta_y = 0.0281$ m.

Assume that the mated assembly was stress free at a temperature of 20 C and is to be heated to a uniform temperature of 300 C. Still in the assembly process, begin a SW Simulation study.

(1) Select the **SW Simulation icon** to open its manager panel. Because the assembly has touching faces the **Component Contacts** feature defaults to **Global Contact** with fully bonded surfaces. If that was not desired you would manually have to edit the **Contact Sets** group. (Later, you will try unbounded or free surfaces for comparison purposes.)

(2) Right click on the **Simulation → New Study**. In the **Study panel** give a **Study name**, pick the static **Analysis type**, click **OK**.

Having given a study name and a mesh type you would usually assign material properties. However, that assignment was made within SolidWorks while constructing the two individual beam parts. You could edit the imported properties at this point if you have decided on a different material.

Remember that the material properties for each beam were selected within SolidWorks. They have been imported into SW Simulation and may need to be reviewed or edited. In this problem you have bonded materials subjected to the same temperature change.

Therefore, the coefficients of thermal expansion (actually their difference) are very important and should be double checked. Assume that has been done. Identify the thermal loading early in the process:

(1) Right click on the **Study name** → **Properties** (rather than the **Temperature** panel).
(2) In the **Static panel** pick **Flow/Thermal Effects**. Check the **Include thermal effects**, and pick **Input temperature**.
(3) Enter the final temperature **value and units** and the **stress free** (zero strain) temperature value of 20 C, click **OK**.

For future reference you should note that the **Static panel** provides for non-uniform temperature distributions to be imported from external files saved in a prior thermal study.

As with any problem, you must prevent the six possible rigid body motions (RBMs). That must be done in a way that does not restrict the thermal motion, unless the true support system does that. You also need to invoke the two symmetry planes used to reduce the computational costs. The first (short) symmetry plane restrains rigid body motion in the direction of the long axis of the beam (i.e., normal to the plane). Since there are multiple points on that restraint plane it will also prevent RBM about the two coordinate axes lying in that plane (so three RBM remain to be addressed).

The second (long) symmetry plane must also prevent displacement normal to it, and thus also prevents the third remaining RBM rotation (1 RMB translation remains). Finally, one point must be restrained to eliminate the remaining possible translational RBM (in the direction parallel to the corner formed by the two symmetry planes). Any point in the assembly could be picked and all computed motions in this last direction will be computed *relative to that chosen point*. The most logical point to chose is in the corner junction of the first two restraint planes, and at the interface of the two materials at the center of the beam. The basic restraint steps are:

(1) Right click on **Fixtures** → **Advanced** → **Symmetry**. Pick the two flat faces at the vertical center. The assembly looks like a cantilever with respect to that plane and forms a half-symmetry

model. Likewise, enforce vertical plane symmetry in the axial direction by picking the two back faces. They give a quarter symmetry model.

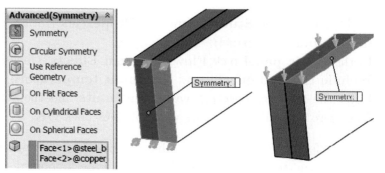

(2) However, one rigid body motion remains to be eliminated. Use **Fixed Geometry** and select the corner vertex as the **Selected Entities**. (Actually, you only needed to fix the displacement parallel to the intersection line of the first two restraint planes because two of the point restraints are already done.)

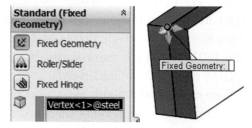

You should expect the temperature change to cause a bonded beam to curve. Therefore, bending stresses will develop. Due to the bonding very large interface shear stresses should also be expected. A crude mesh will probably give a reasonably accurate displacement calculation. However, to get accurate bending stress recovery you need at least 5 to 6 layers of elements through the thickness of each material. That requires mesh control on each of the four faces sharing the thickness dimension. Bending fiber stresses are most likely to be large on the top and bottom layer of *each* material. Thus they need a good surface mesh too. You can either lower the average element size to get that and/or manually control the element sizes. The most rapid changes in the stresses will occur at the free end, and possibly at the center plane. Use **Mesh → Mesh Control** to force finer meshes at the two ends.

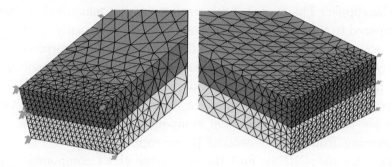

Remember that there are no external mechanical loads to be applied here, only the uniform temperature change. There are many displacement degrees of freedom in this mess and most are wasted since there will be little change in displacements through the width, w. While you should get good deflection results the three elements through each material thickness may not give good stress estimates. (A shell model, in plane stress, probably would have been more accurate, but only in the y–z plane, and had much fewer equations to solve.) All that remains is to input the uniform temperature change for both bodies.

Use **Thermal Loads → Temperature** to reach the **Temperature Panel**. There, expand the part tree so you can see the names of both parts. Pick each part name from that tree and verify that they appear in the selected items list. Set the type to temperature (versus Initial that could be used in a transient model). Enter the uniform temperature change as 300 C.

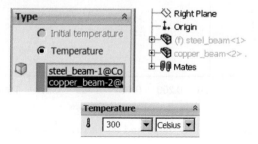

Now start the equation solver with **Study name** → **Run** in the manager panel. A look at the deformed shape shows the shape that was expected due to the higher thermal expansion coefficient of the copper. The overall displacement size (un-magnified) is large compared to the total beam thickness. That would usually mean that you should consider a geometrically non-linear large displacement analysis, but it might not be justified for this thermal loading. If the materials had not been bonded you would have just axial (UZ) displacements. Looking at them in detail, as seen in Figure 14.2, you see that axial displacement component is always positive, but decreasing from top to bottom (from copper to steel). The total displacements (URES), and the transverse (UY) components, are given in Figure 14.2. Both are much larger that the previous axial components. You can create a list of the largest UY components by zooming in on the region and picking nodes with the Probe feature.

The temperature difference not only causes a curvature in the long (axial) direction of the assembly, but also in the narrow direction. That is a three-dimensional structural response that the much more cost effective plane stress approximation would have totally missed. To see that effect (Figure 14.3), create a graph of the vertical direction along the interface edge at the free end. While its value is small, compared to the about 28 mm of end motion, the

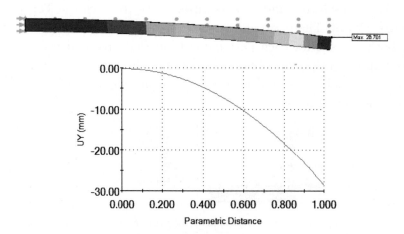

Fig. 14.2.   Deformed shape and its vertical component graph.

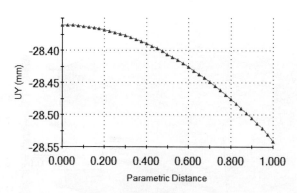

Fig. 14.3. Transverse curving at interface plane, at the free end.

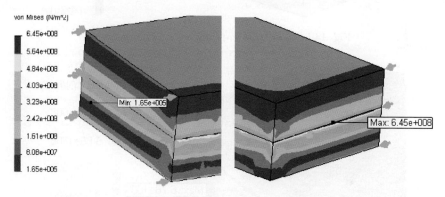

Fig. 14.4. Fixed and free end von Mises stress magnitudes.

occurrence of bowing of the interface along the entire length of the beam is predicted by three-dimensional elasticity theory.

The stresses in the bimetallic beam should be essentially constant except very close to the free end. Thus, you need only display the two ends in detail. The von Mises stresses at each end are given in Figure 14.4. The wiggles in the contours imply that the mesh is too crude near the center bonding plane (which is an important region for this load state).

At this point (because of Figure 14.4) you should already expect that this mesh is too crude to rely upon for stress results. To verify that, consider the free end region yz-shear stress and the maximum 3D shear stress plot of Figure 14.5, and Figure 14.6, respectively.

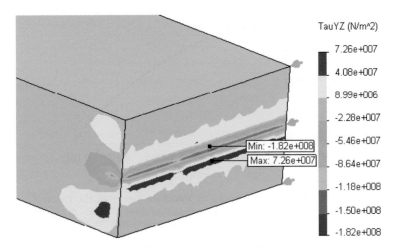

Fig. 14.5. Free end yz-stress component values.

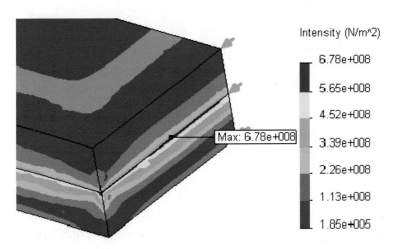

Fig. 14.6. Maximum 3D shear stress at the free end region.

Clearly, the contour lines have many wiggles and the study should be repeated with a finer mesh and/or a plane stress mesh. If you look at the deflection and stress plots in all three coordinate planes (not shown here) you see that this 3D model has mainly a 2D response. Very little is changing through the 40 mm half width. That also suggests using a shell model for the first phase of a problem like this one. After doing that you could more wisely control a 3D mesh for a final study.

Wiggles in stress contours can also be caused by using the default small deflection theory (as done here) to obtain a result that seems to be a large displacement result (greater than half the smallest material or part thickness). The above results suggest that this study should be rechecked, after the final part revisions, with a large displacement analysis option (see below).

It may be surprising how complex the end stress regions are for this simple geometry. It is common for the end region of one material bonded to another to develop a weak stress singularity. Keep that fact in mind when designing initial meshes so that you do not miss important results by having large elements that average out the local details to the point that you cannot see them.

While you may not trust the current stress results, you should still compare the computed displacements (which are always more accurate that the computed stress) to the initial engineering estimate of the maximum deflection. The computed maximum magnitude, in Figure 14.2, was about 2.87 mm. The original deflection estimate was $\delta_Y = 2.81$ mm, so the agreement was surprisingly good.

A large deflection analysis requires an iterative loading and displacement solution loop. In each loop a small part of the (thermal) loading is added to form a new resultant load. The corresponding displacements are computed and added to the original positions of the elements' nodes. The updated element positions are employed to re-compute the element stiffness (conduction) matrices. The looping is repeated about 10 to 20 times to obtain the final displacements and the last set of element stresses. That is called a *"geometrically non-linear analysis"* (as contrasted to a *"materially non-linear analysis"* which is much more complicated). If it does not fail, a large deflection analysis runs about 50−60 times longer that the standard analysis presented above. This was done, but the process resulted in only minor changes. The large deflection study mainly reinforced the observation was that this problem requires a much finer mesh, even though the part is geometrically quite simple.

Fig. 14.7.   Deflection of the layers when the interface is not bonded.

The bonding makes a great difference in the response of any assembly. To see that change, you could modify this assembly, and omit the bond at the touching surfaces as shown in Fig. 14.7. The resulting stresses should be zero and the displacements will only be in the UZ axial direction (try it).

14.3. Ram Block Thermal-Stress Study

The previous example was simple enough that you could reasonably assume that it was subjected to a uniform temperature change (as a steady state result). The other, more common class of thermal stress problem requires the non-uniform temperatures to be computed before the stress study and to be imported into the stress model. It is not unusual for a totally different program to be used for that purpose. Then there is a potential problem in those results being from an incompatible mesh. If so, it is necessary to project or interpolate the external temperature data onto the stress mesh nodes. That is a potential source of error, so check the transfer carefully if you have to use such a procedure.

In SW Simulation you can avoid that problem simply by doing the thermal study first. Then you open the following thermal-stress study the software defaults to using the output from as the input to the stress study. That file selection is under user control, so you could import thermal results from another source.

Here, the previously discussed Ram Block thermal study results will be input to determine the resulting thermal stress state. Those stresses will be compared to the even earlier Ram Block pressure loading stress results. To do that re-open the Ram Block model and create a thermal stress study. Copy the part, restraints, materials, and mesh into the new study. Suppress the pressure loading. It could be un-suppressed later to see the combined effects of temperatures and pressures.

The only new action required is to import the mesh temperature files:

(1) Right click on the **Thermal stress part name** → **Properties** → **Flow/Thermal** Effects and check the default temperature results **file name** for the input temperatures (Ram_block).

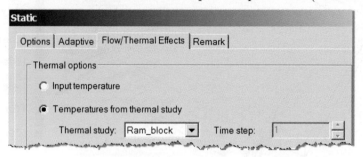

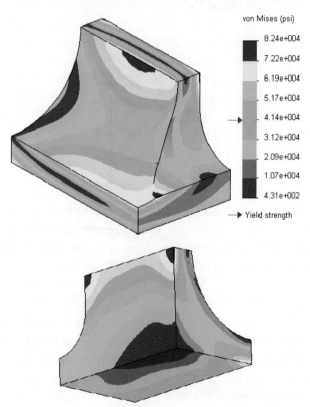

Fig. 14.8. Huge thermally induced von Mises stresses.

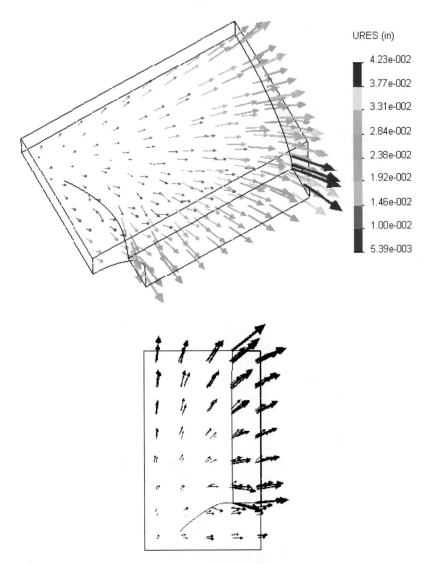

Fig. 14.9.   Ram Block thermally induced displacements.

(2) You must also define the reference temperatures for which the part is stress free. Use a room temperature value, say 77 F, **OK**.

| Reference temperature at zero strain: | 76.73 | Fahrenheit | ▼ |

Next, you simply run the study and check the results. The von Mises stresses due to the temperatures alone are shown in Figure 14.8. They are huge, about twice the yield stress level of the material, and about twice the level due to the pressure loading (Figure 5.17). Clearly, there is no need to evaluate the combination of pressure and temperature loads. The part must be completely re-designed. It would be wise to begin an optimization study to vary the parametric dimensions for the wall thicknesses. For completeness the thermal displacements are in Figure 14.9, for comparison to the pressure only displacements in Figure 5.15.

15

Related Analogies

15.1. Basic Concepts

The differential equation used in a finite element study in one discipline often appears in a different discipline, but with a different physical meaning for the unknown and the coefficients in the equation. That is particularly true for the diffusion equation (heat transfer here) and the biharmonic equation (flat plate deflection here). They are the most common second order and fourth order differential equations, respectively, in engineering. Therefore, a finite element solution presented for one field of application can often be utilized in another field by analogy. Consider the slightly generalized 2D field equation, in the solution domain:

$$k_x \frac{\partial^2 \varphi}{\partial x^2} + k_y \frac{\partial^2 \varphi}{\partial y^2} + P = 0.$$

Subject to a Dirichlet boundary condition on boundary segment Γ_D of $\varphi = \varphi_{given}$ or a Neumann boundary condition (known normal flux) on Γ_N of

$$k_n \frac{\partial \varphi}{\partial n} = f_{given}$$

or a convection (Robin) boundary condition on boundary segment Γ_R of

$$k_n \frac{\partial \varphi}{\partial n} = h(\varphi - \varphi_\infty) = h\varphi + g,$$

where the boundary segments do not overlap. The meanings of the above symbols, in a few disciplines, are listed in Table 15.1. These

Table 15.1. Some general field equation terms.

φ	k	P	h	φ_∞	f
Temperature	Thermal conductivity	Heat power	Convection coefficient	Convection temperature	Boundary flux
Voltage	Electrical conductivity	Current source	0	0	Boundary current
Hydraulic head	Soil permeability	Flow source	0	0	Boundary flow
Velocity potential	1	Flow source	0	0	Boundary velocity

analogies allow you to use SW Simulation to solve problems in such fields by replacing the SW Simulation inputs with corresponding values (and units) for the field of interest. You should also edit the graphic outputs to show the desired terminology (as done here with the SnagIt software).

15.2. Seepage Under a Dam

Water in a reservoir will almost always seep under and/or around a dam. It needs to be controlled to have a low velocity so that the region around the dam will not be eroded away. This seepage, or porous media flow, field will be illustrated for a dam resting on layered, and thus orthotropic, soils [13]. Here, φ is the hydraulic or piezometric head, measured in *meters of water*. The P term is a source or sink of water flow (injection or removal well, m^3/day) and is zero in this example. The fluid velocity components through the soil are defined by Darcy's law, which is directly analogous to Fourier's law (for orthotropic soil in material directions): $V_x = -k_x \partial\varphi/\partial x$, $V_y = -k_y \partial\varphi/\partial y$.

The dimensions of the soil regions, dam, and toe wall are given in Figure 15.1 in meters. The left side of the dam holds water 30 m deep, while the right side is 1 m deep. Split lines locate the impervious dam interface at the soil top. The far boundaries of the soil are also impervious (no normal flow, $f = 0$, the natural boundary condition). The layer soil permeabilities (or hydraulic conductivities)

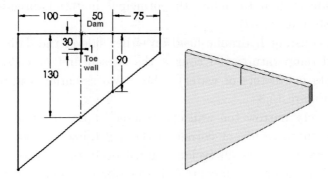

Fig. 15.1. Orthotropic soil beneath a dam.

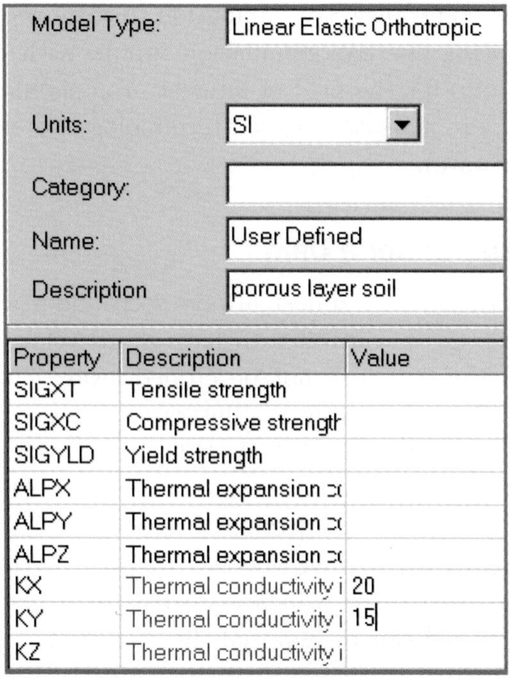

Model Type:	Linear Elastic Orthotropic
Units:	SI ▼
Category:	
Name:	User Defined
Description	porous layer soil

Property	Description	Value
SIGXT	Tensile strength	
SIGXC	Compressive strength	
SIGYLD	Yield strength	
ALPX	Thermal expansion x	
ALPY	Thermal expansion x	
ALPZ	Thermal expansion x	
KX	Thermal conductivity i	20
KY	Thermal conductivity i	15
KZ	Thermal conductivity i	

Fig. 15.2.   Orthotropic soil permeabilities in the Front Plane.

are 20 m/day and 15 m/day in the horizontal and vertical directions, respectively. Those two orthotropic properties are specified with respect to the Front Plane and input as seen in Figure 15.2. Usually layered soils are inclined and require the use of a local material coordinate system to define the principal material directions (as previously illustrated).

The constant hydraulic head on either side of the dam are like specified temperatures. Of course, the chosen unit of *Celsius* here represents m (meters of water). Those two essential restraints are seen in Figure 15.3.

The very narrow toe wall constructed at the front of the dam causes a sharp reentrant corner in the soil (almost a crack). That means very high gradients (velocities) will occur there. Therefore, it is necessary to invoke mesh control there to force small elements at the base of that wall. A portion of the soil mesh is given in Figure 15.4.

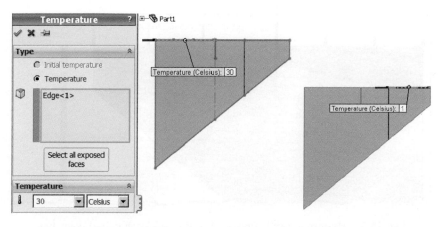

Fig. 15.3.   Set lake (left) and stream water pressure boundary values.

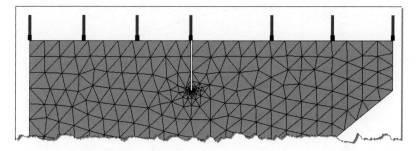

Fig. 15.4.   Refined mesh around the toe wall tip.

Now you can **Run** the study. The resulting soil water pressures
are given in Figure 15.5. The image was captured and then edited to
label the color bar for the current units. The purpose of the toe wall
is to lower the uplift pressure under the gravity dam. The graph of
the pressure along the base of the dam is in Figure 15.6. It would
have varied from 30 to 1 m without the toe wall. Note that it has a
very steep gradient at the downstream end of the base of the dam.
The slope of that graph is related to the speed of the flow.

In this analogy, the heat flux magnitude corresponds to the speed
of the water through the soil. The flow speed is shown in Figure 15.7.
As expected, the re-entrant corners cause a local singularity at the
base of the toe wall. There, the water moving down the left side of

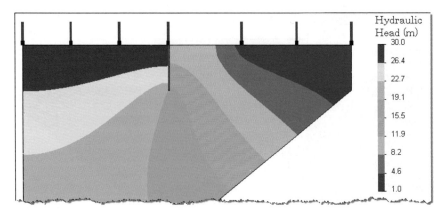

Fig. 15.5.   Hydraulic head (pressure) in the soil, with edited color bar.

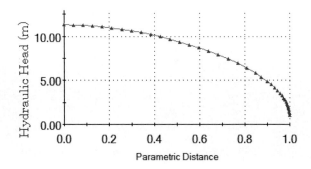

Fig. 15.6.   Pressure along the base of the dam.

the wall rapidly reverses direction as it moves up the right side of the wall. A second singularity at the downstream edge of the dam is also noted.

That is due to another type of singularity not discussed before. There you have an essential boundary condition to the right of the point but a no flux boundary condition to the left of the point (edge of the dam). A mathematical discontinuity occurs there. Its presence was not accounted for in the original mesh. A refined mesh gives new speed estimates in Figure 15.8.

To see the velocity vectors, just select the SW Simulation heat flux vector plot, copy it, and re-label the color bar to display the units of m/day (with SnagIt, etc.). As desired, the velocities are quite small through the soil. The largest values occur where the water

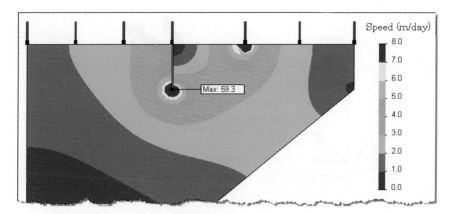

Fig. 15.7. Speed of the ground water flow.

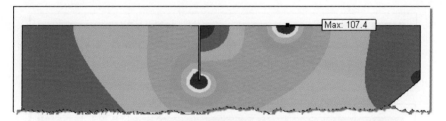

Fig. 15.8. Revised speed for refined mesh at downstream dam point.

changes directions from down to up around the toe wall and at the downstream edge of the dam (see Figure 15.9). The solution segments without the toe wall are seen in Figure 15.10 (using the same contour levels as above). There a high water flow level under the entire base of the dam is observed.

15.3. Potential Flow Around a Cylinder

The flow of an inviscid fluid can be formulated in terms of either a velocity potential or a stream function. The former works in both 2D and 3D implementations, so it is selected for this example. The governing differential equation is

$$\rho \left(\frac{\partial^2 \varphi}{\partial x^2} + \frac{\partial^2 \varphi}{\partial y^2} \right) + Q(x, y) = 0,$$

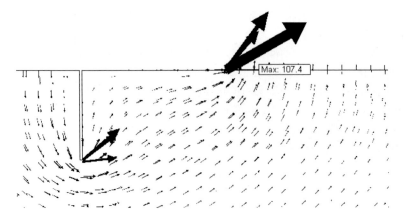

Fig. 15.9.   Seepage velocity vectors at the toe wall tip.

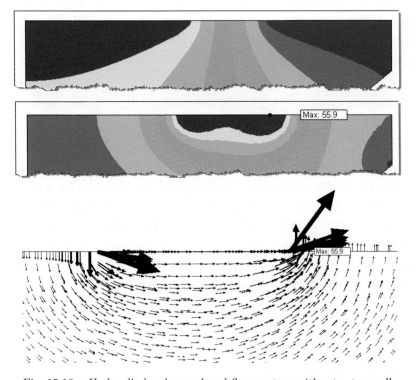

Fig. 15.10.   Hydraulic head, speed and flow vectors without a toe wall.

where Q is a source or sink of mass flow per unit area, ρ is the mass density and φ is the velocity potential. Therefore, potential flow can be solved with a heat transfer code by giving the inputs and outputs a different interpretation. Usually, $Q = 0$, and the density is constant so the equation reduces to the Laplace equation. The velocity vector components are the gradient components of the potential: $u = \partial\varphi/\partial x$, $v = \partial\varphi/\partial y$. The normal inlet or outlet flow is usually specified on the boundary as

$$u_n = \vec{\nabla}\varphi \cdot \vec{n} = \frac{\partial\varphi}{\partial x}n_x + \frac{\partial\varphi}{\partial y}n_y \ .$$

Consider the Irrotational flow of an ideal fluid around solid cylinder within a rectangular channel dimensioned. The material property of thermal conductivity represents the mass density and is simply set to unity as given in that figure. The top and bottom edges and the cylinder boundaries default to no normal flow (insulated in the thermal sense). The fluid enters at the left with a constant normal velocity of $u = 5\,\mathrm{cm/s}$ (and $n_x = -1$), and exits at the right with the same speed in order to conserve mass. Only those two equal and opposite Neumann boundary conditions are required in theory to determine the value of the velocity potential to within an arbitrary constant. In practice, due to machine accuracy slightly violating mass conservation, you must pick one point to assign an arbitrary value to the potential (to pick the arbitrary constant). Here, use the top wall point centered over the cylinder and set the constant to zero.

For potential flow, a velocity inward across a boundary is negative. The sources must satisfy conservation of mass. Since the length of the outlet is the same as the inlet only the sign changes at the right end outflow. Those two flow loads are illustrated in Figure 15.11. The mesh, in Figure 15.12, was controlled to be finer where the velocities are expected to change rapidly around the cylinder.

The primary unknown, velocity potential, does not have a physical meaning but its value shown in Figure 15.13 confirms the expected anti-symmetric distribution. The velocity magnitudes and vectors are given in Figure 15.14, respectively.

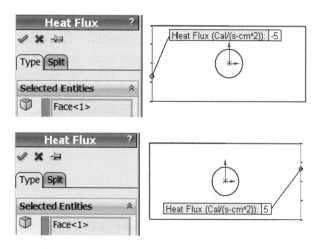

Fig. 15.11.　Inflow and outflow boundary restraint (cm/s).

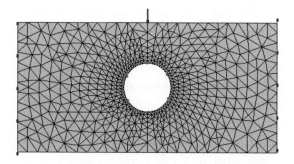

Fig. 15.12.　Graded mesh around the cylinder.

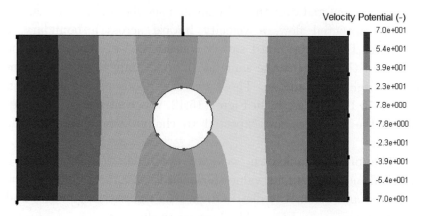

Fig. 15.13.　Anti-symmetric velocity potential around the cylinder.

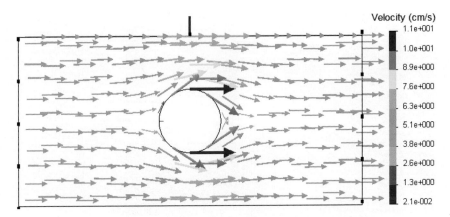

Fig. 15.14.   Velocity vectors with anti-symmetric boundary flows.

Be alert for analogies that can extend the power and usefulness of your finite element software. Many commercial systems offer specific input and output interfaces for the alternate disciplines, but the underlying numerical calculations are basically the same. Minor exceptions are the torsional analogy and the pressurized membrane analogy which both utilize the integral of the solution additional as partial output.

References

1. Adams, V. and A. Askenazi, *Building Better Products with Finite Element Analysis*. 1999, Santa Fe: Onword Press.
2. Akin, J.E., *Finite Element Analysis with Error Estimators*. 2005, Amsterdam: Elsevier/Butterworth-Heinemann.
3. Blevins, R.D., *Formulas for Natural Frequency and Mode Shape*. 1979, Malabar, FL: Krieger.
4. Carslaw, H.S. and J.C. Jaeger, *Conduction of Heat in Solids*. 1959, Oxford: Oxford Press.
5. Chapman, A.J., *Fundamentals of Heat Transfer*. 1987, Collier Macmillan.
6. Jiji, L.M., *Heat Transfer Essentials: A Textbook*. 1998, Begell House Publishers, Inc.
7. Kurowski, P.M., *Engineering Analysis with SW Simulation Professional 2006*. 2006, SDC Publishing.
8. Myers, G.E., *Analytical Methods in Conduction Heat Transfer*. 1971, New York: McGraw-Hill.
9. Norton, R.L., *Machine Design: An Integrated Approach*. 2006, Prentice-Hall.
10. Oden, J.T. and E.A. Ripperger, *Mechanics of Elastic Structures, 2nd Edition*. 1981, New York: McGraw-Hill.
11. Pilkey, W.D., *Formulas for Stress, Strain, and Structural Matrices*. 1993, New York: John Wiley and Sons, Inc.
12. Popov, E.P., *Engineering Mechanics of Solids*. 1990, Prentice-Hall, Inc.
13. Segerlind, L.J., *Applied Finite Element Analysis*. 1984, New York: John Wiley.
14. Timoshenko, S. and S. Woinowsky-Krieger, *Theory of Plates and Shells*. 1987, New York: McGraw-Hill.
15. Young, W.C. and R.G. Budynas, *Roark and Young on TK*. 2002, Universal Technical Systems Rockford, IL.
16. Young, W.C., R.G. Budynas and R.J. Roark, *Roark's Formulas for Stress and Strain*. 2003, McGraw Hill.
17. Young, W.C. and R.G. Budynas, *Roark's Formulas on Excel*. 2005, Universal Technical Systems, Rockford, IL.
18. Ziegler, H., *Principles of Structural Stability*. 1968, Blaisdell.

References

Index